SDN-Supported Edge-Cloud Interplay for Next Generation Internet of Things

SDN-Supported Edge-Cloud Interplay for Next Generation Internet of Things is an invaluable resource covering a wide range of research directions in the field of edge-cloud computing, SDN, and IoT. The integration of SDN in edge-cloud interplay is a promising framework for enhancing QoS for complex IoT-driven applications. The interplay between cloud and edge solves some of the major challenges that arise in traditional IoT architecture. This book is a starting point for those involved in this research domain and explores a range of significant issues including network congestion, traffic management, latency, QoS, scalability, security, and controller placement problems.

Features:

- The book covers emerging trends, issues and solutions in edge-cloud interplay.
- It highlights research advances in SDN, edge, and IoT architecture for smart cities, and software-defined internet of vehicles.
- It includes detailed discussion of performance evaluations of SDN controllers, scalable software-defined edge computing, and AI for edge computing.
- Application areas include machine learning and deep learning in SDN-supported edge-cloud systems.
- Different use cases covered include smart health care, smart city, internet of drones, etc.

This book is designed for scientific communities including graduate students, academics, and industry professionals who are interested in exploring technologies related to the internet of things, such as cloud, SDN, edge, internet of drones, etc.

Chapman & Hall/CRC Internet of Things: Data-Centric Intelligent Computing, Informatics, and Communication

The interconnected roles of adaptation, machine learning, computational intelligence, and data analytics are becoming increasingly essential in the IoT systems field. IoT-based smart systems generate a large amount of data that cannot be processed by traditional data-processing algorithms and applications. The growing capabilities of intelligent systems depend upon various self-decision-making algorithms in IoT devices. This book series covers the various different computational methods used in the IoT-enabled environment, in particular analytics reasoning, learning methods, artificial intelligence, and sense-making in big data.

This series is aimed at researchers and practitioners working in information technology and computer and data sciences, in particular in intelligent computing paradigms, big data, machine learning, sensor data, and the internet of things. The main aim of the series is to make available a range of books on all aspects of learning, analytics and advanced intelligent systems and related technologies. The series covers the theory, research, development, and applications of learning, computational analytics, data processing, and machine learning algorithms, as embedded in the fields of engineering, computer science, and information technology.

Series Editors:

Souvik Pal
Sister Nivedita University, (Techno India Group), Kolkata, India

Dac-Nhuong Le
Haiphong University, Vietnam

Security of Internet of Things Nodes: Challenges, Attacks, and Countermeasures
Chinmay Chakraborty, Sree Ranjani Rajendran and Muhammad Habib Ur Rehman

Cancer Prediction for Industrial IoT 4.0: A Machine Learning Perspective
Meenu Gupta, Rachna Jain, Arun Solanki and Fadi Al-Turjman

Cloud IoT Systems for Smart Agricultural Engineering
Saravanan Krishnan, J Bruce Ralphin Rose, NR Rajalakshmi, N Narayanan Prasanth

Data Science for Effective Healthcare Systems
Hari Singh, Ravindara Bhatt, Prateek Thakral and Dinesh Chander Verma

Internet of Things and Data Mining for Modern Engineering and Healthcare Applications
Ankan Bhattacharya, Bappadittya Roy, Samarendra Nath Sur, Saurav Mallik and Subhasis Dasgupta

Energy Harvesting: Enabling IoT Transformations
Deepti Agarwal, Kimmi Verma and Shabana Urooj

SDN-Supported Edge-Cloud Interplay for Next Generation Internet of Things
Kshira Sagar Sahoo, Arun Solanki, Sambit Kumar Mishra, Bibhudatta Sahoo and Anand Nayyar

SDN-Supported Edge-Cloud Interplay for Next Generation Internet of Things

Edited by
Kshira Sagar Sahoo
Arun Solanki
Sambit Kumar Mishra
Bibhudatta Sahoo
Anand Nayyar

CRC Press
Taylor & Francis Group
Boca Raton London New York

CRC Press is an imprint of the
Taylor & Francis Group, an **informa** business

A CHAPMAN & HALL BOOK

ISBN: 978-1-032-10149-1 (hbk)
ISBN: 978-1-032-39865-5 (pbk)
ISBN: 978-1-003-21387-1 (ebk)

DOI: 10.1201/9781003213871

Typeset in Palatino
by SPi Technologies India Pvt Ltd (Straive)

Contents

v

Preface

With significant advances in communications technology, the internet of things (IoT) today plays a key role in a variety of fields, including transportation, energy, healthcare, and more. In this system, billions of sensor devices are connected to the global network infrastructure and generate enormous amounts of data, which places a huge burden on traditional IoT architectures. In traditional IoT infrastructures, any detection data needs to be sent to a remote storage cloud for further processing. However, remote cloud systems in remote locations are unable to provide for the massive amount of network traffic generated, as well as satisfactory quality of service and quality of experience. Edge computing is a powerful extension of cloud computing, and the interplay between cloud computing and edge computing is an important research area. The edge computing paradigm brings computing and data storage closer to the devices where it is collected, rather than relying on a remote cloud that may be located far away; thus real-time data does not experience latency issues that can affect application performance. The main challenges are network attenuation and congestion through the communication channel. Competent technology is needed that can make well-informed decisions about redirecting traffic to accessible computer end nodes.

5G technology along with a software-defined network (SDN) provides better data transmission mechanisms. SDN provides many benefits including reduced costs, ease of implementation and management, better scalability, availability, flexibility and granular traffic control, etc. SDN integrated within the edge-cloud interplay enhances QoS for complex IoT-driven applications. But various challenges remain, such as lack of security, energy management, controller placement management, and distribution strategies. This book covers cutting-edge methodologies, key technologies, use cases, review documents, and innovative use of SDNs for the interaction between edge and cloud in next-generation IoT infrastructures.

Chapter 1 discusses the detailed SDN-based IoT architecture with edge computing which offers a successful and robust environment for dealing with billions of devices in a meaningful and secure manner. A critical view of current trends in managing IoT devices and the various challenges related to privacy and security issues are highlighted. Further, the advantages of SDN-based IoT implementation are discussed, with conclusive evidence from real-time applications. Chapter 2 provides a brief overview of the evolution, capabilities, and applications of cloud-edge concepts, examining issues and challenges, along with use cases, real-time applications and future research trends. IoT edge devices generate a remarkable volume of raw data, leading to the evolution of accurate design models and many other smart applications in place of the network edge.

Edge computing (EC) along with software-defined network (SDN) and artificial intelligence (AI) has helped to resolve various problems including security challenges, resource allocation, task offloading, improving decision making, etc. SDN simplifies the complexity of IoT networks and AI helps with data privacy. Chapter 3 provides an insight into SDN-aided EC, AI-enabled IoT computing, and how technologies such as these have proved beneficial for the development of IoT devices and smart cities. By segregating the control and data planes, SDN facilitates network management, and SDN architecture can be used in data centers, wide area networks, vehicular ad-hoc networks, named data networks and 5G networks. SDN can address major issues in vehicular communications through network function virtualization (NFV) and centralized controls. Chapter 4 presents a comprehensive overview of software-defined internet of vehicles, its different functionalities, and a layered architecture which improves the quality of service in vehicular communications.

The key feature of a smart city is to provide sustainable livelihood to urban populations by utilizing smart information sharing and developed infrastructure, although the actual deployment of smart cities has been hampered by the need to manage massive amounts of data, and by the platform incompatibilities of linked smart items. With this in mind, Chapter 5 considers whether the current trend in IoT is feasible and sufficient to develop an ideal smart city. The key component of SDN is the programmed network controller consisting of the various applications that run on top of a virtual machine in an IoT network to support various services for IoT applications. Chapter 6 compares several state-of-the-art schemes for performance evaluation of SDN controllers, which is of great value for network engineers, end users and research communities. Several comparison schemes combine features and performance (hybrid approach).

An edge computing model known as fog computing equipped with SDN-enabled fog devices is introduced in Chapter 7. This can process a large number of latency-sensitive IoT applications as well as resource-intensive IoT applications. Wearable sensors and vision-based approaches are common components of conventional health information systems. In this connection, the authors of Chapter 8 explore two case studies related to deep learning and the IoT, as well as the contribution of SDN to making healthcare smarter and much more efficient. Chapter 9 discusses management of IoT traffic using various SDN applications.

We hope that readers will enjoy this book and gain new insights into areas of SDN, IoT and edge computing. The book has great potential to be adapted as a textbook for various courses in cloud computing.

About the Editors

Kshira Sagar Sahoo received his MTech degree in information and communication technology from Indian Institute of Technology Kharagpur, Kharagpur, India, in 2014, and a PhD in computer science and engineering from the National Institute of Technology Rourkela, Rourkela, India, in 2019. He is currently working as Assistant Professor with the Department of CSE, SRM University, Amaravati, AP, India. He is a postdoctoral fellow with the Department of Computing Science, Umea° University, Umea°, Sweden. He has published more than 90 research articles in various top international journals and conferences including *IEEE TITS, IEEE TNSM, IEEE System Journal, IEEE IoT Journal, ACM TOMM, FGCS Elsevier*, and *JSS Elsevier*. His research interests include future-generation network infrastructure, such as SDN, edge computing, the IoT, and industrial IoT. He has more than five years' teaching experience, two years of industry experience, and four years of research experience. He is a member of the IEEE Computer Society and an associate member of the Institute of Engineers (IE), India.

Arun Solanki received his MTech degree in computer engineering from YMCA University, Faridabad, Haryana, India, and a PhD in computer science and engineering from Gautam Buddha University in 2014. He is currently working as an assistant professor in the Department of Computer Science and Engineering, Gautam Buddha University, Greater Noida, India. He is also co-convener of the Center of Excellence in Artificial Intelligence. Dr Solanki has worked as timetable coordinator, and as a member of the examinations, admissions, sports council, digital information, and other university committees. He has supervised more than 70 MTech dissertations and is currently guiding five students through their PhD in artificial intelligence. His research interests span expert systems, machine learning, and search engines. He has published more than 70 research articles in SCI/Scopus-indexed international journals/conferences, and participated in person in many national and international conferences, chairing many sessions. He has been a technical and advisory committee member of many international conferences and has organized several FDPs, conferences, workshops, and seminars. He is an associate editor for the IGI Scopus-indexed *International Journal of Web-Based Learning and Teaching Technologies* (IJWLTT). He has been a guest editor for special issues of *Recent Patents on Computer Science* (Bentham Science Publishers). Arun Solanki has published more than ten books with reputed publishers including IGI Global, CRC, Elsevier, and AAP. He acts as a reviewer for Springer, Wiley, MDPI, IGI Global, Elsevier, and other reputed publishers of SCI/Scopus journals.

Sambit Kumar Mishra is currently working as an assistant professor in the Department of Computer Science and Engineering, SRM University, India. He received his PhD in computer science and engineering from the National Institute of Technology, Rourkela, India, and MTech and MSc degrees in computer science from Utkal University, India. His research interests include cloud computing, edge/fog computing, the internet of things, and wireless sensor networks. He has published more than 50 research articles in internationally reputed journals and conferences. He is a member of the IEEE Computer Society, IETE, and InSc.

Bibhudatta Sahoo obtained his MTech and PhD degrees in computer science and engineering from NIT Rourkela, India. He is presently associate professor in the Department of Computer Science and Engineering, NIT Rourkela, and he has 25 years' teaching experience at undergraduate and graduate level in the field of computer science and engineering. He has authored or co-authored over 200 publications in refereed international journals and conferences by Wiley, Springer, and Elsevier, including *IEEE Transactions*. His technical interests include data structures and algorithm design, parallel and distributed systems, networks, computational machines, algorithms for VLSI design, performance evaluation methods and modeling techniques, distributed computing systems, networking algorithms, and web engineering. He is a member of IEEE and ACM.

Anand Nayyar received his PhD in computer science (wireless sensor networks, swarm intelligence and network simulation) from Desh Bhagat University in 2017. He is currently working in the School of Computer Science-Duy Tan University, Da Nang, Vietnam as Assistant Professor, Scientist, Vice-Chairman (Research) and Director (IoT and Intelligent Systems Lab). He holds 125 professional certificates from CISCO, Microsoft, Amazon, the EC, Oracle, Google, Beingcert, EXIN, GAQM, Cyberoam and many more. He has published more than 150 research papers in various high-quality ISI-SCI/SCIE/SSCI impact factor journals/ Scopus/ESCI-indexed journals, 100 papers in international conferences indexed with Springer, IEEE Xplore and ACM Digital Library, more than 50 book chapters in various Scopus/Web of Science-indexed books with Springer, CRC Press, Wiley, IET and Elsevier. He is a senior or life member of more than 50 associations, including IEEE and ACM. He has authored, co-authored or edited 40 books on computer science. He has 18 Australian patents, 11 Indian Design-cum-Utility patents, three Indian copyrights, two Canadian copyrights, four German patents and one US Patent to his credit in the areas of wireless communications, artificial intelligence, cloud computing, IoT, and image processing. He has won many awards for his teaching and research. He acts as associate editor for *Wireless Networks* (Springer), *Computer Communications* (Elsevier), *International Journal of Sensor*

Networks (IJSNET) (Inderscience), *Frontiers in Computer Science, PeerJ Computer Science, Human Centric Computing and Information Sciences* (HCIS), *IET-Quantum Communications, IET Wireless Sensor Systems, IET Networks, IJDST, IJISP, IJCINI, IJGC*. He is Editor-in-Chief of the USA IGI Global journal *International Journal of Smart Vehicles and Smart Transportation (IJSVST)*. He has reviewed more than 3000 articles for various Web of Science-indexed journals. He is currently researching wireless sensor networks, IoT, swarm intelligence, cloud computing, artificial intelligence, drones, blockchain, cyber security, network simulation, and wireless communications.

List of Contributors

Jehad Ali
Ajou University
Suwon, South Korea

Anchal
DCSA
New Delhi, India

Sreenivasa Rao Annaluri
VNR Vignan Jyothi Institute of
Engineering and Technology
Telangana, India

Hemant Kumar Apat
National Institute of Technology
Rourkela, India

Venkata Ramana Attili
Sreenidhi Institute of Science and
Technology
Telangana, India

Balkishan
DCSA
New Delhi, India

Prasant Kumar Dash
C.V. Raman Global
University
Bhubaneswar, India

Shabeg Singh Gill
Indraprastha Institute of Information
Technology(IIIT)
New Delhi, India

Preeti Gulia
DCSA
New Delhi, India

Lopamudra Hota
National Institute of Technology
Rourkela, India

Sahib Khan
University of Engineering and Technology
Mardan, Pakistan

Sudhakar Kumar
Chandigarh College of Engineering and
Technology
Chandigarh, India

Pooja Mittal
DCSA
New Delhi, India

D. Mohan
Sreenidhi Institute of Science and
Technology
Hyderabad, India

Sagarika Mohanty
National Institute of Technology
Rourkela, India

Byeong-hee Roh
Ajou University
Suwon, South Korea

Asit Sahoo
National Institute of Technology Rourkela
Rourkela, Odisha, India

Bibhudatta Sahoo
National Institute of Technology
Rourkela, India

Kshira Sagar Sahoo
SRM University
Andhra Pradesh, India
Umea° University
Umea° 901 87, Sweden

Shreeya Swagatika Sahoo
Siksha 'O' Anusandhan Deemed to be
    University
Odisha, India

Rashandeep Singh
Chandigarh College of Engineering and
    Technology
Chandigarh, India

Srishty Singh
KIIT University
Bhubaneswar, India

Sunil Kr. Singh
Chandigarh College of Engineering and
    Technology
Chandigarh, India

Vedaant Singh
KIIT University
Bhubaneswar, India

1

SDN-Based IoT with Edge Computing

Sreenivasa Rao Annaluri

VNR Vignan Jyothi Institute of Engineering and Technology, Telangana, India

Venkata Ramana Attili and D. Mohan

Sreenidhi Institute of Science and Technology, Telangana, India

CONTENTS

DOI: 10.1201/9781003213871-1

1.1 Introduction

The rapid growth in technology and reducing cost of internet data is allowing most cities to adopt smart facilities. A massive boost for the internet of things (IoT) has allowed engineers to introduce ubiquitous intelligent connectivity and smartness for residents (Lin et al. 2017). The devices involved in IoT provide for most human activities but without human intervention, using instead different types of sensors, actuators, etc. (Alaa et al. 2017).

The introduction of wireless communications, cloud computing techniques, and other digital electronic systems has allowed the exponential growth of IoT appliances. Software-defined networks (SDN) and network functions virtualization (NFV) help to abstract and define network functions (Li & Chen, 2015). They are installed, controlled, and manipulated remotely using various software running on standardized computer nodes of a network. According to Cisco Systems, global machine-to-machine (M2M) connections have increased massively and are projected to grow by 14.7 billion by 2023 (Cisco Systems 2020). An estimated 48% of networking devices of all sorts may be connected with IoT applications. With the introduction of 5G devices, edge computing techniques and network facilities, market demand for IoT applications/appliances for building smart cities may see rapid growth. A tremendous amount of internet data is required for most of the smart-home devices, electronic gadgets, sensing mechanisms, etc. It is difficult to imagine the complexity of network components, connectivity, and related issues. However, the potential risks of cyberattacks and the role of intruders with malicious programs cannot be ignored (Aldowah et al. 2018). Such attacks may cause serious damage to the economy and to reputations.

Network softwarization of SDN and the NFV cloud played a vital role in the telecoms industries by establishing flexibility, manageability, and dynamics (Varadharajan & Tupakula 2014; Yan et al. 2018). Cloud computing techniques are capable of providing a secured cloud environment and protection mechanism. Recently, cloud computing techniques seem to have become more dynamic and flexible in providing network security and dealing with most IoT security threats (Li et al. 2018; da Silva et al. 2016). However, network engineers continue to face critical problems relating to security for cloud-based IoT networks, including bottleneck issues and lack of collaboration. With the rapid growth of IoT devices, it is becoming more challenging for a network operator to define suitable defense mechanisms against the variety of cyberattack types on IoT networks (Kakiz et al. 2017; Phan et al. 2017).

Edge computing has recently made its mark in SDN-based IoT networks due to its open IT architecture. This architecture decentralizes processing power, smoothly enabling mobile computing and different IoT activities. This in turn helps to connect the IoT activities of the network to the remote cloud (Sittón-Candanedo et al. 2019). A new trend of cloud applications has been set for end users using decentralized edge computing rather than the centralized computation methods of the past (Taherizadeh et al. 2018). With the arrival of 5G fast wireless networks along with different types of artificial intelligence (AI) techniques and machine learning tools, computation and analytics have reached a level where the massive data created by SDN networks are analyzed at a faster rate. The way data is handled, processed, and delivered using edge computing has transformed millions of devices around the world. The explosive growth of internet-based devices using IoT effectively for a wide range of real-time applications is helping to drive edge computing systems. This represents a massive boost for different types of real-time applications such as self-driving cars, robotics, automation, etc.

Organization of Chapter

The Chapter is organized as: Section 1.2 gives overview of Edge Computing, SDN and IoT Devices, followed by integration of SDN with Edge Computing in Section 1.3. Section 1.4 stresses on Integration of SDN with IoT Devices. Section 1.5 stresses on security issues with Edge Computing. Section 1.6 highlights future challenges of Edge Computing followed by types of scalability in section 1.7. Section 1.8 gives overview of Networking and section 1.9 concludes the chapter.

1.2 Edge Computing, SDN and IoT Devices

1.2.1 Edge Computing

Since Amazon introduced its Elastic Compute Cloud (Amazon 2020) to expand its business and make various products easily accessible to its end users, the cloud computing concept has gained the attention of different business sectors. This arrangement helps provide centralized flexible resources with a pay-as-you-go cost model for most of its customers. Cloud computing with different service models (such as software as a service (SaaS), infrastructure as a service (IaaS), and platform as a service (PaaS)) has helped most customers with respect to performance, cost and convenience (Tiwari & Joshi 2016). As a result, most service enterprises have shifted to the cloud environment, which also attracts a huge number of end users for daily activities. Rapid development has been witnessed in cloud computing with radical changes in the internet and its speed, as well as in overall usage of mobile phones, tablets, laptops, and other electronic gadgets functioning alongside the internet. These devices have seen massive exponential growth in the market due to their usage in different sectors such as health, education, and smart cities.

Compared with cloud platforms, these smart devices and the sensors connected to them have limited resources, such as memory, battery lifetime, CPU processing speed, etc. The result has been the development of mobile cloud computing (MCC) (Qi & Gani 2012),which has delivered great results and powerful computing capabilities. However, MCC suffers from unpredictable network latency issues due to increasing technology and emerging applications. The main issues in this area seem to be due to the communication latency between mobile devices and the cloud environments. For example, the decision-making time of an auto-driving device should be a few milliseconds. But the MCC takes longer to transmit the large amount of data collected from sensors to the cloud environment for processing. This delay may cause accidents. The new computing paradigm of edge computing has emerged to deal with these challenges. Initially called mobile edge computing or multiaccess edge computing (MEC) (Patel et al. 2014), it has largely overlapping principles and application scopes. The fundamental concept of this computing paradigm is to deploy possible resources, known as edge nodes, on the edge of a network. These edge nodes are placed in close proximity to other edge nodes to utilize their capabilities collectively and help to reduce network latency. Such arrangements also save bandwidth and improve the security and privacy of the network.

1.2.2 Software-defined Networks

High-speed modern internet connectivity has allowed the digital age to take a big leap in modern society towards great improvements in daily life. The internet functions on the principles of packet-based switching and distributed architecture. Traditional IP networks are becoming more complex and harder to manage with expensive equipment (Nencioni et al. 2016). Most of the issues are based on the design of network elements of a distributed network architecture. In general, a traditional IP router consists of a data plane and control plane. The data plane forwards the network packets at high speed, and the control plane implements configuration and management functions which governs the forwarding plane to route the packets. The data plane functions locally, and the control plane implements protocols and distributed algorithms to implement certain services effectively. In simple terms, traditional routers represent a complex networking component with challenging configuration procedures, and incur high operating expenses (OPEX) and capital expenditures (CAPEX).

To overcome these shortcomings and to facilitate the massive needs of the cloud computing infrastructure to meet the requirements of social media network giants, such as Facebook and Google, SDN was introduced, decoupling the forwarding plane from the control plane (Casado et al. 2007). The SDN switch contains a pipeline-based packet forwarding engine along with a simple agent to communicate with a centralized SDN controller. The central controller manages all forwarding rules remotely and also implements the defined network management functions. The replacement of distributed algorithms with centralized algorithms helps to simplify the control plane. In SDN, network innovation activities have risen sharply due to the involvement of multiple open-source controllers (Yan et al. 2015). These controllers create an abstraction layer, which helps to underlie the network switching devices. Network programming mobility in SDN running on top of the controller is developed using network programming languages (Monsanto et al. 2012).

The next-generation SDN aims to offer operators complete control over their networks to manage, configure and program the network functionalities. The signature OpenFlow is being replaced by a set of new interfaces such as P4Runtime, gNMI/OpenConfig, and gNOI (Wang et al. 2019). With the next-generation SDN it is possible to avoid vendor lock-in of data planes, enabling easy integration of traditional networks with SDN devices, and offering a migration path for traditional IP networks. However, NFV virtualizes different types of network services (like load balancing and firewalls) (Dezhabad & Sharifian 2018), which run on proprietary (i.e., dedicated) hardware. This is a complementary to SDN and helps to accelerate network innovation. It allows automation and programmability by shifting to software-based platforms. Central SDN controllers can easily manage VNFs.

1.2.3 IoT Devices

Internet of things (IoT) devices include a variety of wireless sensors, computing accessories and actuators connected wirelessly with a network and have the ability to transmit data (Ray 2016). Some may be non-standard devices but they have wireless networks and transfer huge amounts of data. IoT devices are mostly embedded in other electronic devices or equipment for industrial and medical applications. Examples of IoT devices include smartphones, refrigerators, watches, door locks, fire alarms, etc.

The role of IoT is crucial in present daily living and its impact on various domains ranges from large industrial sectors to tiny wearable devices. Different types of IoT frameworks are available in the market to produce a variety of applications with different objectives (Ammar et al. 2018). These frameworks consider certain guidelines, protocols, rules, and standards to proceed with IoT applications.

The ecosystem characteristics of an IoT framework define the success rate of an application, in which security mechanisms and privacy are pivotal.

1.3 Integration of SDN with Edge Computing

A simple cloud-based IoT network with three important levels – edge computing, fog computing, and cloud computing – is shown in Figure 1.1. The key layers of an IoT system are perception, distribution network, and application.

1.3.1 Edge Computing Level

Edge computing is one of the important aspects of distributed computing topology, where the information processing is kept near the edge (Figure 1.2), so that people who produce

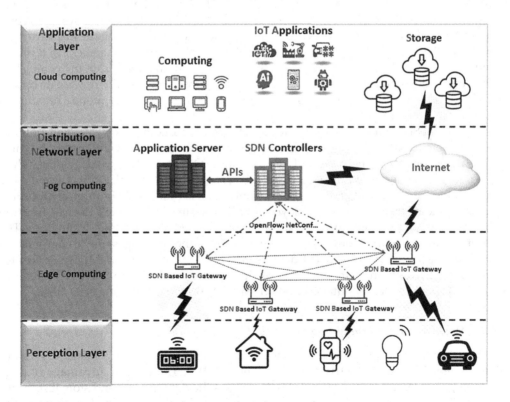

FIGURE 1.1
Environment of a software-defined networks-based IoT network.

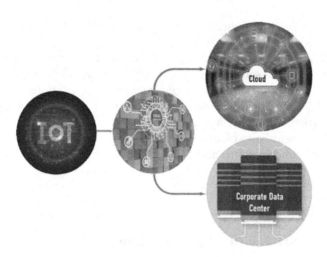

FIGURE 1.2
Functioning of edge computing.

or consume the information have easy access. All the edge nodes such as routers, switches, and small base stations are covered in this level. These nodes are powered by mobile edge computing (MEC) technology, which helps to extend cloud computing by getting it nearer to the edge of networks, leveraging the maximum number of mobile base stations (Abbas et al. 2017). MEC is standardized in the ETSI[1] Industry Specification Group (ISG) (Hu et al. 2015). This has been recognized by 5G PPP (5G Public Private Partnership) as one of the emerging technologies suitable for 5G networks. It works on virtualized platforms to complement NFV which is generally focused on network functions. Hosting the infrastructure of MEC and NFV is quite similar and benefits the infrastructure management in terms of its reusability. The MEC environment is characterized by high bandwidth, proximity, low latency, and radio network information. These are the key characteristics that create opportunity and value for mobile operators by allowing them to monetize mobile broadband experience in a better way.

In the case of SDN-based clouds, the IoT gateways support OpenFlow and NetConf protocols and help to connect with the fog computing level. There are several protocols which establish the connection between IoT devices and SDN-based IoT gateways, including ZigBee, RUBEE, Ethernet, WirelessHART, Wi-Fi, etc. (Al-Fuqaha et al. 2015). For most IoT devices in the perception layer, the real-time connection is established by edge computing.[1] Level from the distribution network layer of an IoT system. All the IoT devices function with the leveraging capacity of MEC technology using edge computing techniques. The computational capability of devices at the edge is one of the challenges due to computation offloading (Mach & Becvar 2017). IoT devices may need a variety of computing services for intensive applications from mobile edge sections and sometimes the massive number of these IoT devices creates an outage situation (Mach & Becvar 2016; Dong et al. 2020).

1.3.1.1 Why Does Edge Computing Matter in SDN Networks?

For most companies, edge computing saves on costs, and is the driving element for maintaining huge data and infrastructure (Baktir et al. 2017). The biggest advantage of edge computing in SDN networks is the ability to process the data quickly and store it securely. For example, facial recognition using a smartphone and a cloud-based algorithm used to take a long time to process. With the edge computing model, the same process can be run locally on the edge servers or gateways, and sometimes even on the same smartphones. Enhanced interconnectivity helps to improve edge access for a maximum number of core applications of an SDN network. The edge infrastructure is compatible with and almost similar to NFV, helping the IoT and telecoms industry to expand with new industry-specific use cases (Baktir et al. 2017; Hsieh et al. 2018). Some of the advantages and benefits of edge computing are listed below:

- High resolution and effective control
- Flexible and low barrier to innovation
- Service-centric implementation
- Virtual machine mobility is possible
- Greater adaptability
- Provides low-cost solutions and interoperability is possible
- Multiplicity of scope is expected

1.3.2 Fog Computing Level

This is a system-level architecture which helps to bring computing, storage and networking functionalities closer to some of the data-producing sources along the cloud range. Introduced by Cisco, it is regarded as an extension of cloud computing to the edge network (Tao et al. 2019). Some of the latest smart devices are capable of processing the data locally, rather than sending it to the cloud, promoting a distributed computing paradigm.

The fog computing level is located in the distribution network layer, which consists of SDN controllers and application servers. Southbound protocols such as OpenFlow and NetConf are used to establish communication between SDN controllers and SDN-based IoT gateways. For the purpose of data exchange, APIs are used with the application servers and also at the level of cloud computing. Therefore, the role of fog computing is critical to provide computational resources along with low latency and help with computation-intensive applications (Escamilla-Ambrosio et al. 2018; Shafi et al. 2018). Due to these qualities, the fog computing level is the most appropriate place to deploy most IoT security applications.

1.3.3 Cloud Computing Level

Cloud computing is involved with the delivery of major computing services, such as servers, storage, databases, software installations, intelligence, etc. over the internet to offer a faster resources management system, in which the network provider's investment is more flexible and economic. It facilitates the access of multiple applications and data from worldwide networking locations using internet connections. It helps to save the overall cost of a business entity and offers scalable computing resources. One of the standard definitions of cloud computing is given by NIST[2] as "… a model for enabling ubiquitous, convenient, on-demand network access to a shared pool of configurable computing resources that can be rapidly provisioned and released with minimal management effort or service provider interaction" (Mell & Grance 2011, p. 2). Large-scale data analysis is easy with cloud computing, unlike edge computing. Therefore, the key responsibility of the cloud computing level is to store, process and access the data generated by different IoT devices and analyze the same data for further usage.

1.3.4 Interaction between SDN and Edge–Cloud

The role of edge computing is similar to that of cloud computing, but reduced transmission latency and improved speed are its two critical features. Localization is possible with the edge and centralization is possible with cloud applications. Therefore, interaction between the core cloud and the edge is unavoidable. Edge–cloud interactions need optimizing, and both industry and academia are working on these challenges. In recent times, an SDN-based edge–cloud interaction system was proposed by Kaur et al. (2018) to maintain quality of service (QoS) in an industrial IoT (IIoT) environment with no network congestion. In this method, the edge–cloud interaction depends on the middleware which is suitable, flexible, and SDN compatible. OpenFlow switches are needed in the data plane; to manage and schedule the traffic flow in the WAN, a central controller was utilized. Scheduling flows takes place in three phases. In the first phase, edge nodes will classify the flow into two categories, batch processing and stream processing. The selection of these categories is based on bandwidth and latency. The control logic selection takes place in the second phase, depending on the classification results. In the third phase, selected logic execution takes place on the SDN controller to obtain the routing and energy-driven flow scheduling (Wang et al. 2019).

One of the largest manufacturers of routers and switches, CISCO, provides easy-access deployment of software-defined WAN (SD-WAN). This deployment, based on Viptela, provides intelligent WAN (iWAN) for companies with complex deployments (Radcliffe et al. 2019). A variety of edge security challenges can be solved by deploying an SD-WAN security stack. This arrangement is capable of shielding operations at the edge for the entire traffic on the network flow with an excellent security management capability. Apart from security, SD-WAN also provides a balanced user experience.

Attackers can use four traffic profiles: 1) data at rest in the branch and cloud; 2) at the open-network ports connected directly to the internet; 3) at the cloud resources and SaaS applications which are bypassed by centralized security solutions; and 4) using personal devices to enable guest access for local Wi-Fi. For all of these profiles, CISCO SD-WAN provides security using an embedded application-aware firewall at the branch router, and through the VPN for applications in the data center. SD-WAN also provides DNS security and intrusion detection to avoid DOS, malware and phishing attacks. The security stack in the SD-WAN provides web filtering, intrusion detection and prevents the internet infections.

1.4 Integration of SDN with IoT Devices

Most edge devices with heterogeneous characteristics require a dynamic networking architecture capable of dealing with evolving challenges, complexities, protocols, and networking resources. In this section, the role of SDN in controlling IoT devices, resource management, and security and privacy issues are discussed.

1.4.1 Role of SDN In Control of IoT Devices

Wireless sensor networks (WSN) play a key role in enabling most IoT devices, and maximum utilization of SDN can be considered for WSN management. SDN WISE is a scalable SDN solution to manage the WSN, which supports duty cycle and data aggregation for reducing total information shared between a variety of sensor nodes and network controllers (Galluccio et al. 2015). Using this method, sensor nodes are programmable and function like finite state machines. Soft-WSN, proposed by Bera et al. (2016), is based on a specialized SDN controller that supports application-aware service provision. The controller proposed in this method is suitable for device management and topology management as needed by IoT to enhance the flexibility and simplicity of the WSN management.

Recently, Internet of Vehicles (IoV) has become one of the key components of IoT, where multiple vehicles with different types of communication technologies such as LTE, 4G, 5G, etc. are interconnected. They form vehicle-to-vehicle (V2V) communication, through which they can scatter with adjacent vehicles to avoid any kind of congestion. SDN provides better vehicular network management and controlling options for the challenges arising from the heterogeneity of different communication protocols, scalability requirements and with diverse QoS. The SDN-based VANET framework suggested by Ku et al. (2014) addressed most deployment issues, like unbalanced traffic flow, inefficient network issues, and flexibility. Roadside units (RSUs) are equipped with local agents, and they are controlled by SDN controllers remotely. These controllers reserve certain communication channels for dealing with emergency traffic, because central controllers implement policies by defining the rules for RSUs. Later, an SDN-based architecture supporting VANET with edge computing was proposed by Truong et al. (2015), focusing on time-sensitive services (such as safety services). The adoption of SDN techniques appears to resolve most of the network management challenges (topology, physical medium, capabilities, and mobility), despite increasing complexity with VANET and the rise in edge nodes.

The integration of SDN with the IoT has served as a tool for improving flexibility and communication between devices. The SDN-enabled IoT network consists of three layers: a) application layer, b) controller layer and c) infrastructure layer. It is almost impossible to control all IoT devices with a single centralized controller, due to limited processing capacity, scalability and reliability. It is therefore essential to deploy multiple controllers to manage the huge number of IoT devices and applications (Liu et al. 2015; Song et al. 2018). However, convenience and flexibility for users seem to be falling short with current management platforms due to the exponential growth in the number of IoT devices and extensive heterogeneity. Xu et al. (2016a), proposed a software-defined smart home (SDSH) platform with three layers using SDN features. The three layers were a) smart hardware layer, b) controller layer, and c) external service layer. In this arrangement the controller layer can be deployed on physical hardware at the user's home or in the abstract equipment of a cloud. The control layer is key to achieve the intelligence, adaptive control and management of smart devices by shielding the hardware details; it even manages the system resources in a centralized manner with task scheduling.

1.4.2 Role of SDN In IOT Resource Management

Multiple heterogeneous IoT devices and network elements spread out to cover wide areas with a large number of IoT subnetworks need to be executed concurrently. Efficiency, sharing of network resources and sensors among a variety of applications is challenging, and appropriate resource management and provisioning are required to meet end-to-end QoS demands.

SDN-based networks are capable of achieving application-specific quality levels in most of the heterogeneous WSN scenarios. A multinetwork information architecture (MINA) proposed by Qin et al. (2014) consists of a layered IoT SDN controller, which delivers commands to control flow scheduling instead of task-level heterogeneous ad-hoc paths. This controller also engages in network calculus and genetic algorithms for the optimization of existing IoT network elements. The data collection component gathers the information related to the devices and network architectures for IoT multi-networks and stores it in databases. Layered controller components can easily utilize this information, enhancing the efficiency and network resource-sharing capabilities.

1.4.3 Role of SDN In IoT Security and Privacy

There are widespread concerns with respect to security and privacy when dealing with IoT in most widespread networks, as multiple users and competitors with individual goals are involved. To enhance IoT security, Miettinen et al. (2017) proposed an IoT SENTINET security system which automatically identifies the IoT device type and applies SDN-enabled mitigation measures. An SDN-based security gateway allows an easy connection between the local IoT network and the internet. A cloud-enabled IoT security service system identifies the device and assesses its vulnerability. The security gateway functions as a core traffic controller and monitoring component. The SDN controller deploys different types of mitigation measures through the security gateway to reduce the risk of harmful events, including device isolation, user notification, and traffic filtering. However, this method is not scalable for large-scale IoT networks.

To provide local-level security, i.e., to avoid global traffic monitoring, Sahoo et al. (2015) proposed an SDN-based ad-hoc IoT network, which leverages SDN controllers to deal with the authentication of local IoT devices. When a new network device is identified, the SDN controller initiates authentication logic by blocking all the switch ports when a connection is established between a switch and controller. The controller installs the flow rules for the switches only after successful authentication. However, this method has several limitations. The network under protection needs to integrate some of the special nodes, which helps to integrate with the SDN controller, programmable data plane, and legacy interfaces. The architecture used in this method is too generic and not scalable to obtain synchronization and consistency.

1.5 Security Issues with Edge Computing

Introducing a new technology always aims to solve a problem, but opens the doors for others, especially when it is dealing with a greater number of networking devices. Edge computing, which deals with the data at the edge, can be more troublesome, especially when it is dealing with SDN networks, where different devices from different network elements may not be secured in the same way as in a cloud-based system (Caprolu et al. 2019). The likelihood of an increase in the number of IoT devices and network elements is greater in SDN networks, making edge computing more vulnerable to different types of security-related issues (Figure 1.3). These systems need to be secured enough to deal with any kind of network intrusions or security breaches. Data encryption, access-control methods, and VPN tunneling are among useful methods to avoid any kind of security issues in SDN networks.

The cloud environment provides connectivity for different IoT devices and applications, and also consist of distributed computational resources and storage. Potential security vulnerabilities in most cloud environments have been identified by experts. Most network-based attacks may influence and harm cloud-based IoT networks. These attacks can be classified in different ways.

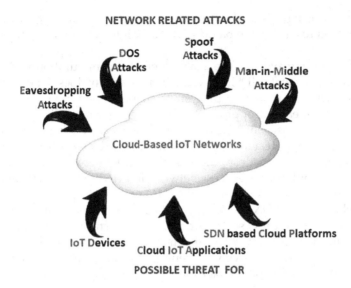

FIGURE 1.3
Attacks and security threats for cloud-based IoT networks.

1.5.1 Eavesdropping Attacks

Data sniffing (also known as eavesdropping attack) is a serious cyberattack carried out by listening to communication between IoT devices. Unencrypted data transfer with sensitive information in unsecured channels is extracted by the sniffers (Xu et al. 2016b). This information could be the credentials or configuration details of a network. Prevention and detection of these type of attacks, also known as network sniffing or network snooping attacks because network connections, components and devices are involved, is a major challenge due to the unchanged mode of network transmission(Hines et al. 2008). The hackers look for a weak connection between different unencrypted network components. At the appropriate time external devices are used or malicious software programs installed in the targets via social engineering. Hackers then intercept data packets traversing the network to read the personal information of the targets or to understand the web traffic on a particular network. Sniffer programs are used by most security teams to monitor and analyze network traffic to identify different types of vulnerabilities and network-related issues. Wireshark, Snort, and TCPDump are some of those sniffing applications, but they are also used by malicious actors to exploit the networks.

Data sniffing attacks can be carried out through mobile phone sensors including cameras, GPS receivers, microphones, etc. (Cai et al. 2009). These types of attacks use side channels that include sound and electromagnetic emanation for the purpose of interfering with keystrokes of a physical keyboard or a smartphone. The hackers use vibrations and motion data from a mobile touch screen to infer which keys were used to type the information, helping them gain valuable information such as passwords, debit/credit card PINs, social security numbers, email passwords, etc. Some intruders use a Trojan program to infiltrate a target computer to log keystrokes. They may also use band channels to infer the keystrokes. Similarly, an acoustic key logger is used to infer keystrokes from acoustic frequency signatures, timing attacks on PINs, and language models (Asonov & Agrawal 2004; Foo Kune & Kim 2010; Cai & Chen 2011). In some cases, accelerometers are being used on a smartphone to infer passwords, and entire sequences of text entered on a touch screen are extracted using this method (Owusu et al. 2012). However, the introduction of touch screens helped to change the paradigm of user interactions as no physical keyboard is involved.

Traditional networking systems use a static network configuration with existing security measures such as firewalls, IDS, IPS, etc. which generally does not prevent eavesdropping. Sometimes network eavesdropping stays silent at the time of the attack, creating confusion. A successful eavesdropping attack has its own preconditions and consists of a stable target path with consistent time duration. This means hackers need to work on the network architecture for quite a long time in order to conduct a successful eavesdropping attack. The probability of eavesdropping attacks in SDN networks is very high. In most SDN-based edge computing, easy-access network information available at the edge soon attracts the attention of hackers. Zhang et al. (2016) proposed a path-hopping communication (PHC) mechanism for SDN, where the ends of the communication path and routing paths are changed dynamically, so that the traffic is distributed in multiple flows and allowed to transmit by different paths. In this way, the network eavesdropping can be potentially successfully and effectively prevented. The following are among best practices to prevent network eavesdropping attacks:

- **Encryption**: This is the most secure way of avoiding eavesdropping attacks. Emails, networks, communication channels, and data in use, rest and motion are encrypted so that hackers have difficulty at the first level. HTTPS is used for web-based communications (Blew et al. 2009), and Wi-Fi protected access 2 or 3 (WPA2/WPA3) based techniques for wireless encryption (Katz 2010).

- **Authentication**: All incoming packets should be authenticated to prevent spoof packets, which are used to execute IP spoofing or MAC address. Standard protocols such as TLS, OpenPGP, secured mail extensions, and IPsec should be used (Callas et al. 2009).

- **Network Monitoring**: Constant monitoring by security teams for any kind of abnormal activities is needed. Some intrusion detection systems are available in the market and in some cases sniffer programs can be used to detect vulnerable attacks on networks.

- **Network Segmentation**: This process keeps a certain amount of data out of reach of hackers. Segmentation of networks will not allow hackers to access all segments of information even if they are successful in one segment.

- **Security Awareness Practices**: Employees must be trained in the risk factors involved in networking infrastructure, data, and components. Strict guidelines must be followed when dealing with network activities. Strong passwords are required and should be changed frequently to avoid networks being accessed by guesswork.

1.5.2 Denial-of-Service Attacks

Denial-of-service (DOS) attacks are the most common and dangerous attacks carried out in cloud-based IoT environments. These attacks tend to shut down a network or a machine completely by flooding the network and IoT devices with huge volumes of traffic. These events tend to exhaust the network quickly or make computational resources unavailable for IoT communication systems. DOS attacks can overload SDN networks easily at the controller processing stage and flood the switch CAM tables to degrade the overall performance of the network (Dridi & Zhani 2016). Similarly, distributed DOS (DDOS) attacks disrupt the traffic of targeted servers, services and networks, crushing targets directly or indirectly by attacking the surrounding infrastructure with a flood of traffic. These attacks are similar to unexpected traffic jams on a highway that prevent regular traffic reaching their destination.

1.5.2.1 Implementation of DDOS Attacks

The main requirement of a DDOS attack is a network with internet-connected machines (such as computers, IoT devices, etc.). These devices (referred to as bots/zombies, with groups of bots termed a botnet) are generally infected with malware, so that the hackers can control all the network components remotely (Singh et al. 2015; Khalaf et al. 2019). In the initial stages, the hackers try to establish a botnet to take control of the network elements by sending instructions to each bot remotely. Once the botnet takes over the victim server, individual bots will send requests to the target's IP address, with the potential to crash the entire network. Countering these attacks is very difficult because all the bots are legitimate internet devices functioning independently according to the random instructions issued by the hackers.

1.5.2.2 Identifying DDOS Attacks

This may sound simple but is very tricky, because suddenly the website under the DDOS attack will become slow or unavailable. Sometimes a spike in traffic on a particular website may be the reason for slow performance. Therefore, traffic analytics tools are helpful to identify the signs of DDOS attacks (Yaar et al. 2003). Some of the known attributes of DDOS attacks are:

- Huge amount of suspicious traffic from a single IP address or range;
- Flood of traffic from similar behavioral profiles (such as browser version, geolocation, type of devices used, etc.);
- A sudden or unexpected surge of requests to a particular endpoint or page;
- Realizing odd traffic patterns that have unnatural spikes at odd hours.

1.5.2.3 Types of DDOS Attacks

There are different types of DDOS attacks that target a variety of components in a network. Most components are vulnerable at different levels of each layer of the network connection. The OSI model of a network consists of seven distinct layers (Figure 1.4). Different types of DDOS attacks or traffic flooding involve target devices or networks unable to resist attacks (Darwish et al. 2013). Different types of attacks at different levels of an OSI model are depicted in Figure 1.4. These attacks are of three types: application layer attacks, protocol attacks, and volumetric attacks.

1) Application layer attacks

The major goal of the hacker in this type of attack is to exhaust a target's resources, creating a denial of service (DOS) in the application layer (sometimes also known as Layer 7 attack) as shown in Figure 1.4. In this layer, web pages are created on a server and delivered for any kind of request as a response to HTTP (Singh et al. 2017; Praseed & Thilagam 2018). This is the place where the hackers target the

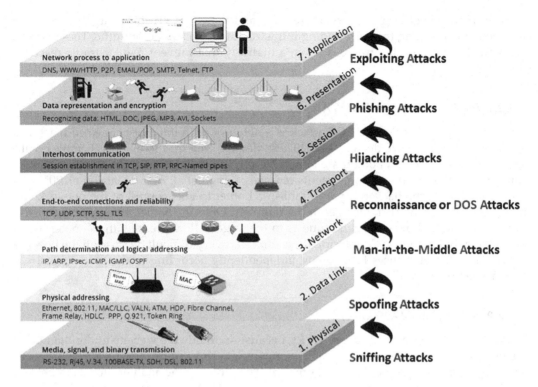

FIGURE 1.4
Types of cyberattacks on different layers of a OSI model.

Source: Adapted from FS Community (2017)

network resources, because sending an HTTP request for computational execution at the client side is cheap. In turn it becomes expensive for the target server to respond to any request, because of multiple files loading and running database queries at the server side.

For example, consider a HTTP flood attack, where different bots are made to refresh the same web browser over and over again (Singh et al. 2017). Such activity creates a huge number of HTTP requests, flooding the server and causing DOS. These attacks may range from simple to complex depending on the number of bots employed by the hacker.

2) Protocol layer attacks

These attacks generate service disruptions by over-consuming server resources or network equipment. Weaknesses of the network layer (layer 3) and transport layer (layer 4) are used to conduct these attacks, making the target inaccessible.

For example, consider a SYN flood, where the target gets requests from bots and goes on receiving the requests from different bots but without obtaining confirmation (Zebari et al. 2018). This process continues until the network cannot carry any more packets and is overwhelmed so that the requests from all the bots remain unanswered. Such attacks exploit the TCP handshake by initiating network connections with the sequence of communications using two computers. In this process, the target receives a large number of TCPs SYN packets with spoof source IP addresses. With the initial connection requests (i.e., TCPs), the target machine is made to respond to each request and wait for the final handshake – which never happens. In this manner, all the target resources are exhausted.

3) Volumetric attacks

These attacks consume all available bandwidth and create congestion between the targets and larger internet resources (Ronen et al. 2019). The target receives a huge amount of data in the form of applications or by some other means from different botnets, which in turn creates massive traffic on the network.

For example, in DNS amplification the Open DNS server receives requests from a spoof IP address, and then the target is made to receive the response from the server itself. In this way, the hackers will generate a long response time with little effort.

1.5.2.4 Mitigating the Process of DDOS Attacks

Differentiating between DDOS attack traffic and normal traffic is challenging. Often a surge in traffic is witnessed, and if it is normal traffic, alleviating efforts can be made to deal with it. DDOS attacks and traffic in the modern internet can come in many forms; they can take the form of a single-source attack or can be a complex attack from different botnets. Complex attacks with adaptive multi-vector DDOS attacks use multiple pathways to overwhelm the target and may also try to distract any mitigation efforts.

For example, in DNS amplification, targeting multiple layers (layers 3 and 4) along with an HTTP flood at layer 7 is quite possible in a DDOS attack (Aizuddin et al. 2017; Singh et al. 2017). A variety of strategies are required to counter and mitigate multi-vector DDOS attacks. Most of these attacks blend the mitigation process as much as possible to make it difficult to separate the normal traffic from the attack traffic (Dimolianis et al. 2019). To deal with such multi-vector DDOS attacks, network engineers either drop or limit the traffic indiscriminately. Sometimes they may even discard good traffic along with the attack traffic. This may also lead to modified attack traffic from hackers seeking to circumvent the counter-measures. Hence, it is always advisable to have a layered solution to obtain the greatest benefit.

- **Blackhole Routing**: A blackhole route is created that funnels the traffic into a narrow route to conduct filtering without any restriction criteria (Khare et al. 2017). In this process, both legitimate and malicious traffic will be routed through a blackhole and then it will be dropped from the network. This method may not be an ideal solution because it is allowing the goal of the hackers – to make the network inaccessible.

- **Rate Limiting**: In this method, the total number of requests for a server will be limited for a certain amount of time to mitigate DDOS attacks (Jiang & Huawei Technologies Co Ltd. 2020). This will allow the web scrapers to slow down data theft and help avoid brute force login attempts. However, effective handling of complex DDOS attacks is not possible due to the involvement of multiple botnets at certain times.

- **Web Application Firewall (WAF)**: This is a tool which helps to mitigate attacks on the application layer (layer 7) by installing a WAF between internet and server (Clincy & Shahriar 2018; Lewis & CenturyLink Intellectual Property LLC 2020). It works as a reverse proxy[3] to protect the target server from any kind of malicious traffic. A series of rules are used for the purpose of filtering and to identify DDOS attacks in layer 7. This method is quite effective due to the ability to adopt custom rules quickly, and provides a quick response to any new attack.

- **Anycast Network Diffusion**: This method uses the Anycast network to scatter the attack traffic of the distributed servers to detect the traffic absorption point in the network (Guleria et al. 2019). It allows distributed attack traffic up to a point and it becomes easier to manage and diffuse any kind of disruptive event. But the reliability of this method depends on the efficiency of the network and the size of the DDOS attack.

1.6 Enhancing SDN for Future Challenges of Edge Computing

SDN has become a driving force in networking technologies, delivering exciting results along with cloud and edge computing. Advances in SDN allow the potential of the edge computing paradigm to be leveraged for various IoT applications. Many advantages can be derived from the hidden scope and potential of SDN and edge computing both independently and in combination. Some of the increasingly sophisticated requirements demand immediate changes. In some networking scenarios, it may not be easy to deploy SDN-enabled networking devices due to some technical issues. Similarly, SDN-enabled hardware is quite expensive to deploy in some places due to lack of adequate investment. Baktir et al. (2017) suggest some of the directions to enhance SDN for edge computing (Figure 1.5). They are of three types: protocols and standardization, scalability, and networking (Singh et al. 2014).

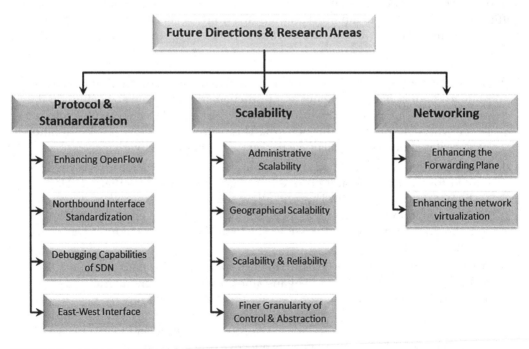

FIGURE 1.5
Some of the ways of enhancing SDN for edge computing.

1.6.1 Protocols and Standardization

Most networking devices need to have upgraded protocols and standards to meet the latest challenges.

1.6.1.1 Enhancing OpenFlow

OpenFlow (v1.5.1) still proves a fundamental solution to hide some complex operations from end users (Baktir et al. 2017). However, many changes are needed to suit traditional network operations and protocols. OpenFlow and SDN need some improvements to support service-centric environments. OpenFlow functions by assuming that forward nodes have fixed behavior, along with a pre-defined set of protocols that supports the network requirements. However, the number of fields using OpenFlow is increasing rapidly, and this in turn prohibits the flexibility of SDN to match schemes as expected. Expectations are high that new matching fields can be implemented with the latest versions of OpenFlow, so that it supports protocol-independent behaviors.

1.6.1.2 Northbound Interface Standardization

The northbound interface also needs to be more responsible and has become important for controlling the underlying networking infrastructure. The role of this interface in SDNs is still raw and challenging due to the multiple interfaces that will be managed in the physical and transport layers (Thyagaturu et al. 2016). It is essential to have a simple vendor-independent interface between network applications and controllers to establish efficient communication. Some of the special APIs are required to cover most of the controllers and an intensive study in this direction is needed to standardize the northbound interface. The standardization process of the northbound interface provides a concrete acceleration for SDN innovation because this interface plays a critical role in control operations.

1.6.1.3 Debugging Capabilities

Both SDN controllers and the northbound applications are software based, and inevitably errors are generated while shifting hardware-based to software-based behavior. In this scenario with edge computing, the realization of different functionalities using the software becomes more complex and prone to maximum errors. It is also difficult to expect control-level cooperation between the cloud and edge computing implementation using the northbound applications due to the hidden bugs within the controller or switch code (Dai et al. 2016). The debugging capabilities of SDN-enabled networks therefore need further improvements.

1.6.1.4 East–west Interface and Geographical Scalability

SDN was previously assumed to be viable for data centers from an infrastructure-as-a-service (IaaS) perspective with a closed campus environment, but with the addition of the edge computing paradigm, the traffic flow must be forwarded through multiple domains. Therefore, the controller of a domain is responsible for establishing the inter-domain communication with adjacent domains (Zhou et al. 2017). East–west interface helps to establish this communication, although SDN is generally not dependent on it. Serious research in this direction is required to establish secure, efficient and reliable communication between the controllers of different networks.

1.7 Scalability

SDN architectures needs to be more flexible and scalable to adopt the latest innovations with the existing infrastructure.

1.7.1 Administrative Scalability

There is a serious problem of overlapping in the virtualization concept due to the flexible nature of edge systems. Some complicated business interactions may also emerge with some of the service models. In this context, SDN is expected to be more flexible to embrace them. With the increase in administrative domains, network environments must be maintained to provide high-end security. Synchronization between different cloudlets also needs to be done carefully to enhance the efficiency of the discovery mechanism (Michel & Julien 2013).

1.7.2 Geographic Scalability

In the case of software-defined optical networks (SDON), east–west interface has attracted the attention of network engineers, due to the ability to transmit communication using multiple transport technologies. Standardization of the interface may help to enhance the quality of operations in SDON to cover larger areas with multiple domains. Metro-scale datacenters need serious consideration; situating servers in metropolitan areas is a challenge, because they have a direct impact on SDN (Sahoo et al. 2016). The distribution of control mechanism is defined by the controller location. Therefore, a scaling of metro-datacenters is needed as per the changes in network modifications.

1.7.3 Scalability and Reliability

In SDN, the central controller may not deliver the necessary ways of controlling the forwarding elements, and scalability may be a serious issue. For example, scaling the OpenFlow with a large networking environment may demand answers for a wide range of challenges, because the SDN operation is based on controller performance, which is centralized logically and distributed physically. It is challenging to achieve distributed controller architecture to enable fully operational high performance (Ros & Ruiz 2014). Therefore, scalability and reliability of SDN to obtain improved controller performance needs detailed research.

1.7.4 Finer Granularity of Control and Abstraction

SDN design proposes a single controller to cover a campus network, but the introduction of edge computing with scope to expand services has led to many reliability problems. Abstraction and finer granularity of control in larger networks plays a vital role. So many transitional challenges were ignored while transforming traditional networks to hybrid SDN-enabled networks (Xu et al. 2017). Additional consideration is needed for abstraction to many hybrid environments where it is more feasible to leverage SDN benefits. For example, most SDN-enabled switches are used for different applications of devices that are not fully OpenFlow, but have the capacity to communicate through OpenFlow. These switches may differ and may lead to additional problems when dealing with multiple domains. Therefore, a barrier is created for SDN integration with the present scenario to obtain finer granularity of control.

1.8 Networking

Networks need to have the ability to enhance their essential functionalities to leverage abstraction and sharing of existing infrastructure.

1.8.1 Enhancing Network Virtualization

Network virtualization leverages the abstraction and sharing of infrastructure to improve overall utilization and resource management. NFV, SDN and network virtualization are inseparable terms due to their inter-beneficial operations. With the involvement of edge computing, the role of these three terms becomes key to the success of various networking applications (Blenk et al. 2016). A detailed study of these platforms covering scalability, performance, abstraction, security and reliability is needed to create a competitive solution for network virtualization. Similarly, the virtualization of wireless networks requires attention due to the involvement of edge computing.

1.8.2 Enhancing the Forward Plane

SDN control plane is becoming more flexible, scalable and resilient with the latest OpenFlow innovations. However, with the addition of edge devices, networks are dealing with more traffic and complexities. Therefore, network performance is at risk due to complex switching requirements. The existing data plane elements lack appropriate CPU power to control operations, because they are not designed for handling larger OpenFlow messages (Kobayashi et al. 2014). In some cases, parallel processing is used to meet the power requirements, or the problem is solved by using powerful servers. Apart from switching capabilities, memory capabilities also play a vital role with an extended number of devices at the edge. They need to support all the extensions of OpenFlow and to avoid any kind of overload problems on the switches due to unexpected attacks.

1.9 Conclusion

This chapter has provided a detailed study of the SDN-based edge computing paradigm to meet the challenging requirements of present market trends with huge numbers of IoT devices and their applications. A simple SDN-based cloud architecture accommodates a wide range of IoT devices at the edge to deliver quality of services for the end users. Various challenges of SDN-enabled networks with the integration of edge and IoT devices has been discussed in detail. Possible security issues with networking attacks are elaborated to understand the performance of SDN-based edge under these challenging scenarios. Most SDN-based networks are quite vulnerable to DDOS attacks, and some of the techniques to prevent possible damage are discussed here.

Notes

1 ETSI (European Telecommunications Standards Institute) is an independent not-for-profit institute, which is known for standardization of the information and communication technology industry.
2 Peter Mell, Timothy Grance, The NIST Definition of Cloud Computing – Recommendations of the National Institute of Standards and Technology, September 2011.
3 Reverse proxy is used to improve the security, performance and reliability of network and it works as a server between web server and clients.

Bibliography

Abbas, N., Zhang, Y., Taherkordi, A. and Skeie, T., 2017. Mobile edge computing: A survey. *IEEE Internet of Things Journal*, 5(1), pp. 450–465.

Alaa, M., Zaidan, A.A., Zaidan, B.B., Talal, M. and Kiah, M.L.M., 2017. A review of smart home applications based on Internet of Things. *Journal of Network and Computer Applications*, 97, pp. 48–65.

Aldowah, H., Rehman, S.U. and Umar, I., 2018, June. Security in internet of things: issues, challenges and solutions. In *International Conference of Reliable Information and Communication Technology* (pp. 396–405). Springer, Cham.

Al-Fuqaha, A., Guizani, M., Mohammadi, M., Aledhari, M. and Ayyash, M., 2015. Internet of things: A survey on enabling technologies, protocols, and applications. *IEEE Communications Surveys & Tutorials*, 17(4), pp. 2347–2376.

Amazon EC2, 2020. Amazon EC2: secure and resizable compute capacity to support virtually any workload. [Online] available at RL: <https://aws.amazon.com/ec2/?ec2-whats-new.sort-by=item.additionalFields.postDateTime&ec2-whats-new.sort-order=desc>, Accessed on September 25, 2020.

Ammar, M., Russello, G. and Crispo, B., 2018. Internet of Things: A survey on the security of IoT frameworks. *Journal of Information Security and Applications*, 38, pp. 8–27.

Asonov, D. and Agrawal, R., 2004, May. Keyboard acoustic emanations. In *IEEE Symposium on Security and Privacy, 2004. Proceedings. 2004* (pp. 3–11). IEEE.

Attili, V.R., Annaluri, S.R. and Srinivas, V.S.P., 2022. *Software-Defined Networking for Future Internet Technology Security Issues in ISDN* (Vol. 1, p. 310). Apple Academic Press, CRC Press, Taylor& Fransis Group.

Aizuddin, A., Atan, M., Norulazmi, M., Noor, M., Akimi, S. and Abidin, Z., 2017. DNS amplification attack detection and mitigation via sFlow with security-centric SDN. In *Proceedings of the 11th International Conference on Ubiquitous Information Management and Communication (IMCOM '17)* (pp. 1–7). Association for Computing Machinery, New York, NY, USA, Article 3. https://doi.org/10.1145/3022227.3022230

Baktir, A.C., Ozgovde, A. and Ersoy, C., 2017. How can edge computing benefit from software-defined networking: A survey, use cases, and future directions. *IEEE Communications Surveys & Tutorials*, 19(4), pp. 2359–2391.

Bera, S., Misra, S., Roy, S.K. and Obaidat, M.S., 2016. Soft-WSN: Software-defined WSN management system for IoT applications. *IEEE Systems Journal*, 12(3), pp. 2074–2081.

Blenk, A., Basta, A., Zerwas, J. and Kellerer, W., 2016. On the placement problem of network virtualization hypervisors for software defined networking.

Blew, E.O., Chang, K.M. and Educational Testing Service, 2009. *Methods for conducting server-side encryption/decryption-on-demand*. U.S. Patent 7,519,810.

Cai, L. and Chen, H., 2011. TouchLogger: Inferring keystrokes on touch screen from smartphone motion. *HotSec*, *11*(2011), p. 9.

Cai, L., Machiraju, S. and Chen, H., 2009, August. Defending against sensor-sniffing attacks on mobile phones. In *Proceedings of the 1st ACM Workshop on Networking, Systems, and Applications for Mobile Handhelds* (pp. 31–36).

Callas, J.D., Price III, W.F. and Allen, D.E., PGP Corp, 2009. *System and method for secure electronic communication in a partially keyless environment*. U.S. Patent 7,640,427.

Caprolu, M., Di Pietro, R., Lombardi, F. and Raponi, S., 2019, July. Edge computing perspectives: architectures, technologies, and open security issues. In *2019 IEEE International Conference on Edge Computing (EDGE)* (pp. 116–123). IEEE.

Casado, M., Freedman, M.J., Pettit, J., Luo, J., McKeown, N. and Shenker, S., 2007. Ethane: Taking control of the enterprise. *ACM SIGCOMM Computer Communication Review*, *37*(4), pp. 1–12.

Cisco Systems. 2020. White Paper, March 9. https://www.cisco.com/c/en/us/solutions/collateral/executive-perspectives/annual-internet-report/white-paper-c11-741490.html

Clincy, V. and Shahriar, H., 2018, July. Web application firewall: Network security models and configuration. In *2018 IEEE 42nd Annual Computer Software and Applications Conference (COMPSAC)* (Vol. 1, pp. 835–836). IEEE.

da Silva, A.S., Wickboldt, J.A., Granville, L.Z. and Schaeffer-Filho, A., 2016, April. ATLANTIC: A framework for anomaly traffic detection, classification, and mitigation in SDN. In *NOMS 2016– 2016 IEEE/IFIP Network Operations and Management Symposium* (pp. 27–35). IEEE.

Dai, M., Cheng, G. and Wang, Y., 2016, June. Detecting network topology and packet trajectory with SDN- enabled FPGA Platform. In *Proceedings of the 11th International Conference on Future Internet Technologies* (pp. 7–13).

Darwish, M., Ouda, A. and Capretz, L.F., 2013, June. Cloud-based DDoS attacks and defenses. In *International Conference on Information Society (i-Society 2013)* (pp. 67–71). IEEE.

Dezhabad, N. and Sharifian, S., 2018. Learning-based dynamic scalable load-balanced firewall as a service in network function-virtualized cloud computing environments. *The Journal of Supercomputing*, *74*(7), pp. 3329–3358.

Dimolianis, M., Pavlidis, A., Kalogeras, D. and Maglaris, V., 2019, April. Mitigation of multi-vector network attacks via orchestration of distributed rule placement. In *2019 IFIP/IEEE Symposium on Integrated Network and Service Management (IM)* (pp. 162–170). IEEE.

Dong, X., Li, X., Yue, X. and Xiang, W., 2020. Performance analysis of cooperative NOMA based intelligent mobile edge computing system. *China Communications*, *17*(8), pp. 45–57.

Dridi, L. and Zhani, M.F., 2016, October. SDN-guard: DoS attacks mitigation in SDN networks. In *2016 5th IEEE International Conference on Cloud Networking (Cloudnet)* (pp. 212–217). IEEE.

Escamilla-Ambrosio, P.J., Rodríguez-Mota, A., Aguirre-Anaya, E., Acosta-Bermejo, R. and Salinas-Rosales, M., 2018. Distributing computing in the internet of things: cloud, fog and edge computing overview. In *NEO 2016* (pp. 87–115). Springer, Cham.

Foo Kune, D. and Kim, Y., 2010. Timing attacks on PIN input devices. In *Proceedings of the 17th ACM conference on computer and communications security (CCS '10)* (pp. 678–680). Association for Computing Machinery, New York, NY, USA.

FS Community (2017). TCP/IP vs. OSI: What's the difference between the two models? [Online] available on URL: <https://community.fs.com/blog/tcpip-vs-osi-whats-the-difference-between-the-two-models.html>, Accessed on October 12, 2020.

Galluccio, L., Milardo, S., Morabito, G. and Palazzo, S., 2015, April. SDN-WISE: Design, prototyping and experimentation of a stateful SDN solution for WIreless SEnsor networks. In *2015 IEEE Conference on Computer Communications (INFOCOM)* (pp. 513–521). IEEE.

Guleria, A., Kalra, E. and Gupta, K., 2019, February. Detection and prevention of DoS attacks on network systems. In *2019 International Conference on Machine Learning, Big Data, Cloud and Parallel Computing (COMITCon)* (pp. 544–548). IEEE.

Hines, B., Rasmussen, J., Ryan, J., Kapadia, S. and Brennan, J., 2008. *IBM WebSphere DataPower SOA Appliance Handbook*. Pearson Education.

Hsieh, H.C., Chen, J.L. and Benslimane, A., 2018. 5G virtualized multi-access edge computing plat-form for IoT applications. *Journal of Network and Computer Applications*, 115, pp. 94–102.

Hu, Y.C., Patel, M., Sabella, D., Sprecher, N. and Young, V., 2015. Mobile edge computing—A key technology towards 5G. *ETSI White Paper*, 11(11), pp. 1–16.

Jiang, W. and Huawei Technologies Co Ltd, 2020. *SDN-Based DDoS Attack Prevention Method, Apparatus, and System*. U.S. Patent Application 16/824,036.

Kakiz, M.T., Öztürk, E. and Çavdar, T., 2017, September. A novel SDN-based IoT architecture for big data. In *2017 International Artificial Intelligence and Data Processing Symposium (IDAP)* (pp. 1–5). IEEE.

Katz, F.H., 2010, April. WPA vs. WPA2: Is WPA2 really an improvement on WPA? In *2010 4th Annual Computer Security Conference (CSC 2010)*, Coastal Carolina University, Myrtle Beach, SC.

Kaur, K., Garg, S., Aujla, G.S., Kumar, N., Rodrigues, J.J. and Guizani, M., 2018. Edge computing in the industrial internet of things environment: Software-defined-networks-based edge-cloud interplay. *IEEE Communications Magazine*, 56(2), pp. 44–51.

Khalaf, B.A., Mostafa, S.A., Mustapha, A., Mohammed, M.A. and Abduallah, W.M., 2019. Comprehensive review of artificial intelligence and statistical approaches in distributed denial of service attack and defense methods. *IEEE Access*, 7, pp. 51691–51713.

Khare, A.K., Rana, J.L. and Jain, R.C., 2017. Detection of wormhole, blackhole and DDOS attack in MANET using trust estimation under fuzzy logic methodology. *International Journal of Computer Network and Information Security*, 9(7), p. 29.

Kobayashi, M., Seetharaman, S., Parulkar, G., Appenzeller, G., Little, J., Van Reijendam, J., Weissmann, P. and McKeown, N., 2014. Maturing of OpenFlow and software-defined networking through deployments. *Computer Networks*, 61, pp. 151–175.

Ku, I., Lu, Y., Gerla, M., Gomes, R.L., Ongaro, F. and Cerqueira, E., 2014, June. Towards software-defined VANET: Architecture and services. In *2014 13th Annual Mediterranean ad HOC Networking Workshop (MED- HOC-NET)* (pp. 103–110). IEEE.

Lewis, R.A. and CenturyLink Intellectual Property LLC, 2020. *Method and System for Implementing High Availability (HA) Web Application Firewall (WAF) Functionality*. U.S. Patent Application 16/119,382.

Li, J., Zhao, Z., Li, R. and Zhang, H., 2018. Ai-based two-stage intrusion detection for software defined iot networks. *IEEE Internet of Things Journal*, 6(2), pp. 2093–2102.

Li, Y. and Chen, M., 2015. Software-defined network function virtualization: A survey. *IEEE Access*, 3, pp. 2542–2553.

Lin, J., Yu, W., Zhang, N., Yang, X., Zhang, H. and Zhao, W., 2017. A survey on internet of things: Architecture, enabling technologies, security and privacy, and applications. *IEEE Internet of Things Journal*, 4(5), pp. 1125–1142.

Liu, J., Li, Y., Chen, M., Dong, W. and Jin, D., 2015. Software-defined internet of things for smart urban sensing. *IEEE communications magazine*, 53(9), pp. 55–63.

Mach, P. and Becvar, Z., 2016. Cloud-aware power control for real-time application offloading in mobile edge computing. *Transactions on Emerging Telecommunications Technologies*, 27(5), pp. 648–661.

Mach, P. and Becvar, Z., 2017. Mobile edge computing: A survey on architecture and computation offloading. *IEEE Communications Surveys & Tutorials*, 19(3), pp. 1628–1656.

Mell, P. and Grance, T., 2011. The NIST definition of cloud computing. *NIST Special Publication* 800–145.

Michel, J. and Julien, C., 2013, November. A cloudlet-based proximal discovery service for machine-to-machine applications. In *International Conference on Mobile Computing, Applications, and Services* (pp. 215–232). Springer, Cham.

Miettinen, M., Marchal, S., Hafeez, I., Asokan, N., Sadeghi, A.R. and Tarkoma, S., 2017, June. Iot sentinel: Automated device-type identification for security enforcement in iot. In *2017 IEEE 37th International Conference on Distributed Computing Systems (ICDCS)* (pp. 2177–2184). IEEE.

Monsanto, C., Foster, N., Harrison, R. and Walker, D., 2012. A compiler and run-time system for network programming languages. *ACM Sigplan Notices*, 47(1), pp. 217–230.

Nencioni, G., Helvik, B.E., Gonzalez, A.J., Heegaard, P.E. and Kamisinski, A., 2016, June. Availability modelling of software-defined backbone networks. In *2016 46th Annual IEEE/IFIP International Conference on Dependable Systems and Networks Workshop (DSN-W)* (pp. 105–112). IEEE.

Owusu, E., Han, J., Das, S., Perrig, A. and Zhang, J., 2012, February. Accessory: password inference using accelerometers on smartphones. In *Proceedings of the Twelfth Workshop on Mobile Computing Systems & Applications* (pp. 1–6).

Patel, M., Naughton, B., Chan, C., Sprecher, N., Abeta, S. and Neal, A., 2014. Mobile-edge computing introductory technical white paper. *White Paper, Mobile-Edge Computing (MEC) Industry Initiative*, pp. 1089–7801.

Phan, T.V., Bao, N.K. and Park, M., 2017. Distributed-SOM: A novel performance bottleneck handler for large- sized software-defined networks under flooding attacks. *Journal of Network and Computer Applications*, *91*, pp. 14–25.

Praseed, A. and Thilagam, P.S., 2018. DDoS attacks at the application layer: Challenges and research perspectives for safeguarding Web applications. *IEEE Communications Surveys & Tutorials*, *21*(1), pp. 661–685.

Qi, H. and Gani, A., 2012, May. Research on mobile cloud computing: Review, trend and perspectives. In *2012 Second International Conference on Digital Information and Communication Technology and it's Applications (DICTAP)* (pp. 195–202). IEEE.

Qin, Z., Denker, G., Giannelli, C., Bellavista, P. and Venkatasubramanian, N., 2014, May. A software defined networking architecture for the internet-of-things. In *2014 IEEE Network Operations and Management Symposium (NOMS)* (pp. 1–9). IEEE.

Radcliffe, D., Furey, E. and Blue, J., 2019, December. An SD-WAN Solution Assuring Business Quality VoIP Communication for Home Based Employees. In *2019 International Conference on Smart Applications, Communications and Networking (SmartNets)* (pp. 1–6). IEEE.

Ray, P.P., 2016. A survey on Internet of Things architectures. *Journal of King Saudi University-Computer and Information Sciences* (2018) 30, pp. 291–319.

Ronen, R., Neuvirth-Telem, H., Nahum, S.B., Gabaev, Y., Yanovsky, O., Korsunsky, V., Teller, T. and Shteingart, H. and Microsoft Technology Licensing LLC, 2019. *Detecting volumetric attacks*. U.S. Patent 10,425,443.

Ros, F.J. and Ruiz, P.M., 2014, August. Five nines of southbound reliability in software-defined networks. In *Proceedings of the Third Workshop on Hot Topics in Software Defined Networking* (pp. 31–36).

Sahoo, J., Salahuddin, M.A., Glitho, R., Elbiaze, H. and Ajib, W., 2016. A survey on replica server placement algorithms for content delivery networks. *IEEE Communications Surveys & Tutorials*, *19*(2), pp. 1002–1026.

Sahoo, K.S., Sahoo, B. and Panda, A., 2015, December. A secured SDN framework for IoT. In *2015 International Conference on Man and Machine Interfacing (MAMI)* (pp. 1–4). IEEE.

Shafi, Q., Basit, A., Qaisar, S., Koay, A. and Welch, I., 2018. Fog-assisted SDN controlled framework for enduring anomaly detection in an IoT network. *IEEE Access*, *6*, pp. 73713–73723.

Singh, B., Kumar, K. and Bhandari, A., 2015, October. Simulation study of application layer DDoS attack. In *2015 International Conference on Green Computing and Internet of Things (ICGCIoT)* (pp. 893–898). IEEE.

Singh, D.P., Goudar, R.H., Pant, B. et al., 2014. Cluster head selection by randomness with data recovery in WSN. *CSIT* **2**, 97–107. https://doi.org/10.1007/s40012-014-0049-1

Singh, K., Singh, P. and Kumar, K., 2017. Application layer HTTP-GET flood DDoS attacks: Research landscape and challenges. *Computers & Security*, *65*, pp. 344–372.

Sittón-Candanedo, I., Alonso, R.S., Corchado, J.M., Rodríguez-González, S. and Casado-Vara, R., 2019. A review of edge computing reference architectures and a new global edge proposal. *Future Generation Computer Systems*, *99*, pp. 278–294.

Song, C., Wu, J., Chen, X., Shi, L. and Liu, M., 2018, March. Towards the partitioning problem in software- defined IoT networks for urban sensing. In *2018 IEEE International Conference on Pervasive Computing and Communications (PerCom)* (pp. 1–9). IEEE.

Taherizadeh, S., Jones, A.C., Taylor, I., Zhao, Z. and Stankovski, V., 2018. Monitoring self-adaptive applications within edge computing frameworks: A state-of-the-art review. *Journal of Systems and Software, 136*, pp. 19–38.

Tao, F., Zhang, M. and Nee, A.Y.C., 2019. Chapter 8-Digital Twin and Cloud, Fog, Edge Computing. In *Digital Twin Driven Smart Manufacturing*, pp. 171–181.

Thyagaturu, A.S., Mercian, A., McGarry, M.P., Reisslein, M. and Kellerer, W., 2016. Software defined optical networks (SDONs): A comprehensive survey. *IEEE Communications Surveys & Tutorials, 18*(4), pp. 2738–2786.

Tiwari, P.K. and Joshi, S., 2016. Data Security for Software as a Service. In *Web-Based Services: Concepts, Methodologies, Tools, and Applications* (pp. 864–880). IGI Global.

Truong, N.B., Lee, G.M. and Ghamri-Doudane, Y., 2015, May. Software defined networking-based vehicular ADHOC network with fog computing. In *2015 IFIP/IEEE International Symposium on Integrated Network Management (IM)* (pp. 1202–1207). IEEE.

Varadharajan, V. and Tupakula, U., 2014. Security as a service model for cloud environment. *IEEE Transactions on Network and Service Management, 11*(1), pp. 60–75.

Wang, A., Zha, Z., Guo, Y. and Chen, S., 2019. Software-defined networking enhanced edge computing: A network-centric survey. *Proceedings of the IEEE, 107*(8), pp. 1500–1519.

Xu, H., Li, X.Y., Huang, L., Deng, H., Huang, H. and Wang, H., 2017. Incremental deployment and throughput maximization routing for a hybrid SDN. *IEEE/ACM Transactions on Networking, 25*(3), pp. 1861–1875.

Xu, K., Wang, X., Wei, W., Song, H. and Mao, B., 2016a. Toward software defined smart home. *IEEE Communications Magazine, 54*(5), pp. 116–122.

Xu, Q., Ren, P., Song, H. and Du, Q., 2016b. Security enhancement for IoT communications exposed to eavesdroppers with uncertain locations. *IEEE Access, 4*, pp. 2840–2853.

Yaar, A., Perrig, A. and Song, D., 2003, May. Pi: A path identification mechanism to defend against DDoS attacks. In *2003 Symposium on Security and Privacy, 2003.* (pp. 93–107). IEEE.

Yan, Q., Huang, W., Luo, X., Gong, Q. and Yu, F.R., 2018. A multi-level DDoS mitigation framework for the industrial internet of things. *IEEE Communications Magazine, 56*(2), pp. 30–36.

Yan, Q., Yu, F.R., Gong, Q. and Li, J., 2015. Software-defined networking (SDN) and distributed denial of service (DDoS) attacks in cloud computing environments: A survey, some research issues, and challenges. *IEEE Communications Surveys & Tutorials, 18*(1), pp. 602–622.

Zebari, R.R., Zeebaree, S.R. and Jacksi, K., 2018, October. Impact analysis of HTTP and SYN flood DDoS attacks on apache 2 and IIS 10.0 web servers. In *2018 International Conference on Advanced Science and Engineering (ICOASE)* (pp. 156–161). IEEE.

Zhang, C., Bu, Y. and Zhao, Z., 2016, October. SDN-Based Path Hopping Communication Against Eavesdropping Attack. In *Optical Communication, Optical Fiber Sensors, and Optical Memories for Big Data Storage* (Vol. 10158, p. 101580J). International Society for Optics and Photonics.

Zhou, H., Wu, C., Cheng, Q. and Liu, Q., 2017. SDN-LIRU: A lossless and seamless method for SDN inter- domain route updates. *IEEE/ACM Transactions on Networking, 25*(4), pp. 2473–2483.

2

Towards Cloud-Edge-of-Things
State of the Art, Challenges and Future Trends

Lopamudra Hota

National Institute of Technology, Rourkela, India

Prasant Kumar Dash

C.V. Raman Global University, Bhubaneswar, India

CONTENTS

DOI: 10.1201/9781003213871-2

2.1 Introduction

> Cloud-Computing is the third wave of digital revolution
>
> **Lowell McAdam.**

From industries to business and from education to smart technologies, cloud computing has advanced immensely to provide cost-efficient and time-saving services. Cloud computing (CC) has transformed data storage, access, and processing mechanisms. All companies are seeking to adapt to technological change as rapidly as possible, with the number of devices connected for digitization increasing considerably in recent years [1]. The integration of numerous devices means that the internet of things (IoT) now pervades our mundane lives. For example, public transport systems are interconnected by IoT. Sensors connect vehicles to traffic lights and other vehicles as well as other units to build a smart transportation system whose functionalities also include vehicle management and monitoring, parking information management, driver behavior monitoring and autonomous vehicle management. The integration of IoT and the cloud paradigm has improved life prospects, providing solutions and opportunities for intelligent services.

The incorporation of IoT devices in the cloud has improved the performance, storage, processing and analysis of real-time data gathered from IoT-based sensors [2]. The global COVID-19 pandemic in 2020 promoted the relevance of cloud services for IoT-enabled devices for health and activity monitoring [3]. The new term cloud of things (CoT) is an amalgamation of cloud computing technology and IoT [4].

CC plays a significant role in ubiquitous IoT implementation by wirelessly connecting devices that may be static or moving. Although the cloud provides numerous benefits, including large computation space, resources, pay-per-use services, access from anywhere at any time, and delay-constraint services, it still faces challenges when it comes to service delivery to end users – providing on-time delivery with reliable connectivity and minimized cost is a demanding task. Connectivity between end devices and the cloud can be enhanced by practicing edge computing technology. For better energy efficiency and optimized performance, cloud and edge computing are merged to provide services to mobile-end IoT devices and users [5]. One of the most common features of cloud-edge computing is efficient data offloading from devices with limited resource to edge devices. This mechanism provides efficient video streaming, augmented reality, real-time services, video conferencing with QoS and reduced latency [6].

Edge computing (EC) has played a prominent role in optimal utilization of resources and network overheads, and minimizing response times. Processing capabilities are dragged to the edge of the network for faster and more efficient applications and services to edge devices, removing the focus and dependency on centralized systems. EC provides reduced-latency real-time applications, scalability, low cost of operation, and quality of service (QoS). McKinsey Global Institute calculates that the size of the global economy, influenced by IoT and EC, will reach $11 trillion by 2025 [7]. The EC and CC has provided on-demand services with companies like Apple, Google, Amazon, and Cisco providing computing revolution with technological innovation. In the EC environment, applications perform autonomous operations then transmit via cloud, reducing computational and network overheads, and enhancing privacy and security [8]. EC with IoT capabilities and cloud infrastructure can be integrated to provide high-performance, delay-constrained

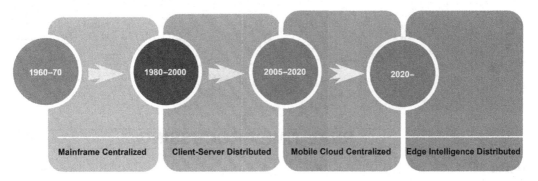

FIGURE 2.1
Evolution of edge-cloud technology.

applications for wireless communications, like mobile ad-hoc networks (MANETs), vehicular ad-hoc networks (VANETs), intelligent transport systems (ITS), and in health care and smart environments. The incorporation of edge computing in the cloud provides globalized support with centralized service. Figure 2.1 illustrates the evolution of edge-cloud technology over time, from mainframe centralized computing to an edge intelligence distributed computing mechanism.

The objectives of this work are: to define cloud-edge computing; explore current standards and architectures for the cloud-edge model; illustrate real-time applications and use cases; and discuss future trends and challenges.

Section 2.2 explores the background and capabilities of cloud-edge computing and compares edge computing with traditional data centers. Sections 2.3 and 2.4 present an overview of cloud and IoT, and edge and IoT, respectively. The challenges faced by edge computing, potential use cases and real-time applications are explored in Sections 2.5, 2.6, and 2.7. Section 2.8 enlists open issues and future trends. Section 2.9 concludes the chapter with future scope.

2.2 Background to Cloud-Edge Computing

The vast domain of cloud-edge computing, with the many applications it offers to developers and services providers at the edge of networks, has many overlapping and conflicting definitions. The main aim is to compute, store and deliver services closer to the input domain or end users. The challenges presented by cloud computing are low and unreliable bandwidth, and latency due to distant centralized cloud data centers. These challenges can be well managed by processing closer to end users, i.e., implementation of edge computing with distributed sites.

The virtualization of network services was the stepping stone to edge computing over wide area networks (WAN) moving away from data centers towards a platform providing flexible, reliable and simple applications with low latency. Cloud-edge computing makes use of mechanisms both for virtualization services and the distributed computing environment. Use cases include industry, IT and education, but also monitoring and tracking in real-time environments including water usage, environmental monitoring and surveillance, etc. These capabilities are provided by IoT gateways or NFV/SDN-based services.

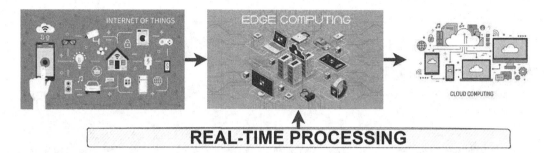

FIGURE 2.2
A view of cloud-edge IoT.

Edge computing is primarily associated with the edge location. As with telecommunications, the edge is a point close to the end user and controlled by the provider, where workloads are running on end-user devices. For large enterprises such as a retail store or factory, the edge is a point where applications, services and workloads are run. Efficient end-user devices capable of supporting IoT and sensor devices and processing the data generated from them are located at the edge [9]. Figure 2.2 illustrates the cloud-edge IoT environment for real-time processing.

2.2.1 Similarities of Edge Computing and Data Center Computing

1. Both edge and data center-based computing have high computation and storage capabilities along with availability of network resources [10].
2. Virtualization and abstraction of the pool of resources simultaneously provides scalable and provisional services to multiple users or clients. The resource pool entails logical abstraction of resources for flexible resource management among multiple groups or clients.
3. The resources are shared by multiple users for executing single or multiple operations or applications.
4. Interoperability using APIs provides seamless connectivity and automated services for enhanced data sharing among various applications.
5. Commodity or off-the-self hardware can be leveraged to reduce costs by providing workstations compatible with different operating systems and environments.

2.2.2 Dissimilarities of Edge Computing and Large Data Centers

1. Edge devices or centers are much closer to the end users, eliminating problems of latency and unreliability [11–12].
2. Edge computing requires specialized hardware like GPU/FPGA and high processing devices for high-end computing like AR/VR.
3. Edge computing is scalable and can scale up to support large sites and distributed locations.
4. Edge computing handles rapid connections and disconnections, supporting a dynamic pool of distributed sites.

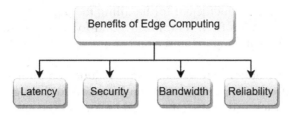

FIGURE 2.3
Benefits of edge computing.

5. Edge sites are resource constrained. Unlike cloud data centers, adding capacity to existing sites is restricted due to insufficient space and power requirements.

6. In some cases, there is a requirement for large-scale multi-tenancy. IoT devices depending on edge computing support multi-tenant operations for virtualization and abstraction.

Note that, any application running on the edge environment leverages the capabilities of the cloud such as storage, computing, virtualization etc.

2.2.3 Capabilities of Edge Computing

Figure 2.3 shows some of the benefits of edge computing:

1. Providing services and performing distributed network services at globally remote locations [13–14].
2. Consistent operating and computing paradigm across varied infrastructures.
3. Integration, orchestration and time-constrained service delivery of applications.
4. Limitations of hardware, cost constraints and bandwidth utilization.
5. Addressing applications with strictly low latency like real-time applications (AR/VR, video streaming, audio services).
6. Security and privacy of data.
7. Geo-fencing data, creating a virtualized geographic boundary using GPS, RFID.

2.3 Overview of Cloud and IoT

As the IoT advances, more and more devices will connect to the internet, including cars and household appliances. The IoT enables connection of smart and autonomous devices with sensors; it provides scalability, reliability and flexibility in varied operations like data processing, storage, accessibility, etc. Cloud-IoT is a groundbreaking innovation that provides efficient connection between the cloud and smart IoT devices with sensors communicating with the cloud via the internet. The IoT devices thus take advantage of cloud resources, including computation, storage for high performance and real-time computing. Due to the enormous rise in the number of devices and data generated (in exabytes per day), cloud technology is inefficient for handling the big data created by IoT devices [15]. Thus, there is a transition from centralized cloud-IoT to decentralized edge services. Among the benefits of combining IoT and cloud are:

1. Many connectivity options are available in IoT cloud computing, implying a large network. Cloud computing resources can be accessed through a wide range of devices, such as mobile devices, tablets and laptops. It is convenient for users but requires network access points.

2. IoT cloud computing can be used on demand by developers. It is, therefore, a web service that can be accessed without special permission. Access to the internet is the only requirement.

3. Users are able to scale the service according to their needs. Having a fast and flexible system allows storage space to be expanded, software settings to be edited, and the number of users to be managed, thus providing deep computing power and storage.

4. Cloud computing means that resources are pooled. As a result, collaboration increases, and users form close bonds.

5. With the growth of IoT devices and automation, security concerns have emerged. Cloud-based authentication and encryption protocols provide companies with re-assurance.

6. Cloud computing with IoT is convenient because what you pay for is exactly what you get. In other words, the cost varies according to usage. Connecting to the internet and sharing data across network components requires an increasingly large network of IP-addressed objects.

Since reliability, security, economy, and performance optimization depend on cloud architecture, it is crucial that it is well designed. A secure environment and agile development can be achieved by using well-designed pipelines, structured services, and sandboxed environments.

The edge model is capable of providing cloud storage and computing capabilities in real time. This is done by moving services from the cloud to edge devices like gateways, PCs, smartphones and other smart devices. This provides high reliability, scalability, low latency, and real-time computational capabilities. The aggregation of edge-cloud and IoT has led to an innovative technology termed the cloud-edge of things.

The need for service delivery to be closer to end-points or end-users defines the need that is driving the development of cloud-edge computing. Edge computing works in conjunction with core capabilities, aiming to deliver enhanced end-user experience without any unreasonable demands on connectivity to the core. This improvement results in reduced latency and mitigating bandwidth limits.

There are trade-offs from the need to greatly increase the number of deployments in order to deliver edge computing facilities, which presents a significant challenge to widespread edge deployment. While managing a single cloud might require a ten-member team, an organization in edge-cloud computing requires many more staff as they need to manage thousands of small clouds. Some of the functionalities to be taken care of are:

1. Standardization and infrastructure consistency.

2. Quick and simple automated management, deployment, failure recovery and replacement.

3. Simple, cost-effective plans for hardware failure.

4. Locally fault-tolerant designs may be important, particularly in environments that are remote or unreachable, so zero-touch infrastructure is desirable. This is a question that balances the cost of buying and running redundant hardware against the cost of outages and emergency repairs.

5. Technicians and other personnel should be trained in automated work like maintenance of software and applications.

6. Other physical parameters like power, humidity and environmental conditions have to be taken into consideration.

2.4 Overview of Edge and IoT

The demand for faster, efficient services and delivery of content on time has motivated organizations to improve edge networks. The huge amount of data generated by the increasing number of connected devices brings with it the challenges of analyzing, managing and storing. Traditionally, these were handled by a private cloud or data center in an organization, but the enormous volume of generated data makes for disadvantages in the traditional approach to storing and processing data.

Edge devices, systems and data centers alleviate the problem of abundant data storage and processing by moving away from centralized access. The storage and processing are nearer to the end-user, reducing latency, cost, efficiency and even low network connectivity.

Unless edge computing is used, the IoT will be reliant on cloud computing and network connectivity. The sending and receiving of data between an IoT device and a cloud can slow response times and decrease efficiency. Other issues that edge computing addresses include network bandwidth requirements for sending huge amounts of data over slow cellular or satellite connections, and allowing systems to continue working when a network connection is lost.

Connected IoT devices generate vast amounts of data through their physical connection that can be leveraged with edge computing. Data is processed locally and used for rapid decision making when analytics algorithms and machine learning models are deployed to the edge. As well as aggregating data before sending it to a central site for processing, edge computing allows data to be maintained for long periods of time. The edge is a physical device with sufficient memory, processing power and computing resources to collect and process data in almost real-time without the assistance of other parts of the network. Edge devices are considered to be those that have enough storage and computation capability to make low-latency decisions and process data within milliseconds. The edge computing environment often implies IoT data, since IoT data are typically generated in a remote location far from a central data center. Over the last decade, the number of smart, internet connected devices has increased tremendously. IoT edge processing capabilities enable businesses to respond faster to new data by deploying them near sensors and devices. Applications include autonomous driving, fleet management and predictive maintenance by gathering data from sensors, cameras and other devices; industrial automation and control in factories to communicate between devices [16]; remote monitoring of oil and gas providing details of flow metrics and pipeline performance; and smart agriculture to improve crop growth, and monitor weather conditions, soil quality, soil moisture, and other essential factors.

2.5 Challenges Faced by Edge Computing

Edge computing platforms must be more fault tolerant and robust than data-center cloud platforms, in terms of both hardware and platform-based application services. Automation and no-touch provisioning in infrastructure and the provided platform are prime requirements for edge computing scenarios.

Challenges of EC include:

1. The implementation cost is high as edge devices are deployed in a distributed manner. The choice of edge devices at various location is a challenge for proficiency in the service provided to edge servers and devices.

2. In a dynamic network architecture framework, computational offloading over edge devices becomes a challenge. The load may tend to increase in some systems, whereas it is reduced in others, which in turn consumes much of the battery power. There is a need to build an energy-efficient system with less overload.

3. EC suffers from network overloading, communication overheads and congestion, requiring an efficient routing protocol for QoS. As the nodes are mobile, mobility management is yet another challenge that degrades network performance.

4. As computation and data are handled at the edge of the network, privacy and security is quite challenging. To handle this issue there must be enhanced trust management and a pseudonym scheme designed to thwart possible attacks and malicious entry.

5. There should be no latency in service delivery, as processing and storage latency deal with real-time services.

Some of these challenges can be overcome by tailoring cloud control for IoT-based services. The cloud mechanism has the functionalities providing scalability, data security, mobility management and disaster recovery for edge devices and communication with IoT.

Edge computing is basically an application running on a local server in order to make the cloud services nearer to the end device. An edge computing architecture consists of local servers to compute capabilities, AI services, with connectivity to IoT or autonomous computing services.

In cloud-edge computing, cloud services are extended, bringing centralized cloud services near to the end-users and processing applications on request at the local, or edge, level, reducing execution time and cost of IoT applications at the edge (Figure 2.4).

2.6 Potential Use Cases

The basic requirements that can be befitted by distributed architecture are security, analytics, compliance and virtualization.

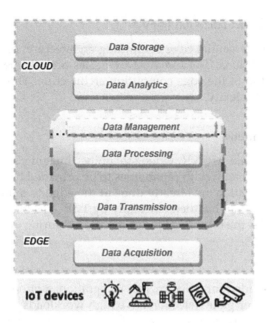

FIGURE 2.4
An architectural view of cloud-edge of things.

2.6.1 Data Gathering and Analytics

The data is collected from various ends, sensors and IoT devices. The data is then sent over limited bandwidth and slow network connections for data analysis to the centralized data center, consuming time and larger bandwidth. Since the volume of data produced is large, taking the computing closer to the source of data generation can be more effective. Data can then be sent to cloud data centers in small chunks. The focus here is balancing the cost of data transmission against data loss.

2.6.2 Security

As edge devices include devices and sensors connected to a network, these are prone to attacks due to the proliferation of end devices. Edge computing has the capability to place the security elements near the source to reduce attacks with high-performance security applications. It also increases the layers of security for potential threats and risk.

2.6.3 Compliance

Compliance includes data sovereignty, copyright, geo-fencing, restricted access to political and geographical boundaries, limitations on data streaming, and storage rules and regulations. These are all achieved and enforced reliably and efficiently with the edge computing infrastructure.

2.6.4 Network Function Virtualization (NFV)

NFV is the heart of quintessential edge computing architecture, providing infrastructure functionalities to applications. The foremost examples are teleoperators looking to run network functions on top of the edge computing architecture. This maximizes efficiency, minimizing cost and complexity.

2.6.5 Immersive Applications

Immersive applications like AR/VR, video streaming, imaging and sensitive healthcare applications require high bandwidth capabilities to be provided by edge computing. Data caching and data optimization are already done at the edge to manage frequent changes in network traffic that are not handled efficiently by TCP. This focus on latency reduction during peak hours varies the bit-rate on radio conditions during video feed.

2.6.6 Efficiency

Many applications like video surveillance, face recognition, motion detection, threat recognition and vehicle number plate detection have high workloads consuming lots of bandwidth. Sending all this information to the cloud wastes network bandwidth, and increases latency and cost. These data can be processed at edge level for anomalies, and further data on which action is to be taken is then reported to the cloud.

2.6.7 Autonomous Operations

Environments with limited and unpredictable connectivity, including transportation, mining, wind and solar power plants, should process data at the edge. Edge computing aids these sites to function semi-autonomously or fully autonomously

2.6.8 Privacy

Personal medical information has to be anonymized before being sent to cloud data centers, and this can be done utilizing edge computing. For cloud-edge computing, privacy of data is maintained at edge servers and then sent to the cloud. Similarly, for companies and organizations, third-party applications are present at the edge infrastructure for privacy then sent to the cloud for storage or computing.

2.6.9 Real Time

Real-time applications, including connected vehicles, healthcare, AR/VR, smart objects, etc. are delay sensitive and are not able tolerate even a few milliseconds of latency [18]. These applications require high bandwidth and low latency, with computation and content caching taking place near the end users. Applications that need high response time and high availability have to be computed and processed by the edge computing infrastructure.

2.7 Real-Time Cloud-Edge Computing Applications

Edge computing plays a vital role in application-oriented commercial, military and industrial services [17–22]. The following are among the use cases.

2.7.1 Autonomous Vehicles or Vehicular Communication

Self-driving vehicles are a prime example of edge computing as they are based on latency-driven applications with large amounts of processing. These data, if sent to cloud data centers for processing, will consume more time, which may lead to dangerous outcomes like accidents. The data need to be processed within a few seconds for a safe driving experience. Computing at the edge level provides real-time decision-making capabilities with reduced latency. Platooning of vehicles is the best use case in this category. Vehicles move in a platoon one behind the other, minimizing congestion and fuel costs. Vehicles here communicate with low frequency, using minimal manpower. All the data and processing can thereafter be stored and sent to cloud data centers. One of the first uses of autonomous vehicles will be the platooning of truck convoys. Several trucks travel together in a convoy, saving fuel and reducing traffic congestion. It will be possible to eliminate the need for drivers in all trucks except the front one due to ultra-low latency communication between the trucks.

Navigational automation is another vehicular communication application. Route information, estimated time and other information en route – such as restaurant availability, petrol pumps, traffic jams, etc. – are updated in real-time. These data need to be stored in edge servers rather than cloud servers for instant updating and rapid access.

Traffic management is yet another area that widely requires edge computing to enable real-time monitoring and management. This includes vehicle optimization, managing autonomous vehicles, lane diversion, roadworks and accidents. These data have to be processed in real-time without delay for immediate decision making.

2.7.2 Oil and Gas Industry

Monitoring assets in the oil and gas sector is essential for disaster management. Oil and gas rigs are in remote locations where it is difficult for networks to handle the data generated. Edge computing can be used for data analytics, providing efficient low-latency processing closer to assets.

2.7.3 Smart Grid

Edge computing can be used in smart grids and aid in better management of energy consumption as the world moves from non-renewable to renewable energy sources. In plants, factories and even office buildings, sensors and IoT devices are connected to the edge for real-time monitoring and analysis of energy consumption. This help to track the amount of electricity consumed at peak times and communicate with the energy provider.

2.7.4 Healthcare

Real-time health monitoring is essential for hospital patients and reduces the burden on doctors. Healthcare has opened up various edge computing opportunities. Sensors can detect patients' temperature, heart rate, blood pressure and movement on a regular basis and these unprocessed data can be stored in the cloud or any big-data platform. Security considerations are important as data is shared via the internet. Installing edge servers on hospital sites could enable local data processing, with the advantage of data privacy, and unusual patient behavior can be tracked via a dashboard. Latency can also be minimized when data are sent within the hospital.

2.7.5 Surveillance

Security cameras record a huge amount of data that is sent to the cloud for processing and storing. Required footage can be kept and that without any movement removed, as this huge amount of data needs high bandwidth to be transmitted to the cloud. Edge devices or the camera itself can preprocess data before sending, reducing bandwidth consumption and network congestion.

2.7.6 Military

Cloud-edge computing can aid rapid computations in a theater of war where lives are at stake. Soldiers in the battlefield needs real-time data for fully efficient combat. Real-time military data fetched and delivered by the internet of military things (IoMT) which includes physical health of soldiers, enemy positions, and arms and ammunition requirements. Data can be sent within the combat arena by deploying rapid-acting edge devices and then sent to the cloud for storage.

2.7.7 Virtualization

Complex processing with low latency is required by the new virtualized RAN hardware, and operators are seeking to combine edge computing with new communication mechanisms for latency-free delivery of services. Some processing and computing needs to be near the cell tower. Parts of the mobile network can be virtualized and computing can be provided at the edge level for cost-efficient low-latency service. Virtualized mobile networks (vRAN) are becoming increasingly popular among operators, as they offer both cost and flexibility advantages.

2.7.8 Content Caching and Delivery

For video streaming, audio transmission or graphic transmission, it is necessary to bring the devices closer to the computing device for low bandwidth and less latency. These can be made possible by edge devices to compute these tasks and deliver services to customers.

2.7.9 Smart Homes, Offices and Buildings

Smart Homes, Offices and Buildings need IoT devices for collecting and disseminating data to be sent to centralized servers for processing and storage [18]. Edge-based computing can bring computing and storage closer to the users, reducing roundtrip time and providing quick delivery with security for sensitive information. For example, voice-based assistant devices like Siri and Alexa respond very rapidly to the user voice command.

2.8 Open Issues and Future Trends

There are many areas of focus for researchers in this field. Software-defined networks (SDN) and NFV, still in their initial phase, have huge potential in the cloud-edge computing paradigm. The implementation of SDN and NFV on cloud-edge IoT platforms will improve reliability and efficient communication. Integrating NFV and SDN is not straightforward, as it involves multiple stakeholders implementing different concepts. The upcoming 6G technology has implications for design and management of networks, enhancing network resilience. IoT devices with intelligent edge computing will be able to communicate with SDN or cloud infrastructure via the 6G network. Data privacy and security is another major challenge due to the large scale of data generated. Among vital research topics are remediation of cloud-edge applications with low cost and convenience; efficient connection and task offloading algorithms for co-operation between mobile devices and cloud-edge; real-time constraint applications with reduced latency and appropriate bandwidth utilization; and intensive computing resource management and handling of offloading task for applications like video gaming, AR/VR and smart devices.

2.9 Conclusion and Future Scope

The dimensions of cloud computing are rapidly evolving towards a future IoT connecting large-scale edge infrastructure with IoT devices. The IoT paradigm generates an enormous amount of data every day, which is difficult for centralized or even decentralized platforms to handle. In this chapter, we have outlined the advances in and benefits of cloud and edge computing, as well as the challenges associated with edge computing. Design of a collaborative cloud-edge model for processing, storage and analysis of data generated by IoT end devices is a research area that is now much in demand. The real-time applications and use cases of cloud-edge computing architecture is also discussed.

Cloud computing will lead ICT infrastructure by 2025, while edge computing will become a rapidly growing market. Although there are many reasons for this, edge computing ultimately supports companies with tasks that cannot be performed in the cloud. With low latency, high data volumes and security, there are some obvious benefits, and neither businesses nor individuals want to be inconvenienced by interruptions in service. As our world moves towards the cloud and edge, distributed cloud and edge architectures will become more important to increase data processing speeds, reduce latency, enable the latest technologies, and sustain the exponential growth of IoT and autonomous things like robots or self-driving vehicles. Companies seeking operational efficiencies should consider moving functionality to the edge in today's data-driven world. In the new ecosystem, hybrid models such as edge-cloud will be promoted by hyperscalers, and Telcos and their service partners will have a significant role to play. The healthcare, manufacturing, retail, transportation, government, and energy industries are all in the process of being connected and digitized. As edge computing and industrial IoT solutions are adopted in the energy sector, a company can increase the performance of large photovoltaic systems, which is advantageous for the grid and the planet. Edge-cloud computing provides numerous opportunities for improving operations and services. It is this, at the forefront of IT decision making and is a fast-growing market.

References

1. https://www.businessinsider.com/internet-of-everything-2015-bi-2014-12
2. Lee, I., & Lee, K. (2015). The Internet of Things (IoT): Applications, investments, and challenges for enterprises. *Business Horizons, 58*(4), 431–440.
3. Hassen, H. B., Ayari, N., & Hamdi, B. (2020). A home hospitalization system based on the Internet of things, Fog computing and cloud computing. *Informatics in Medicine Unlocked, 20*, 100368.
4. Petrolo, R., Loscri, V., & Mitton, N. (2017). Towards a smart city based on cloud of things, a survey on the smart city vision and paradigms. *Transactions on Emerging Telecommunications Technologies, 28*(1), e2931.
5. Mishra, S. K., Sahoo, S., Sahoo, B., & Jena, S. K. (2020). Energy-efficient service allocation techniques in cloud: A survey. *IETE Technical Review, 37*(4), 339–352.
6. Yi, S., Li, C., & Li, Q. (2015, June). A survey of fog computing: concepts, applications and issues. In *Proceedings of the 2015 workshop on mobile big data* (pp. 37–42).
7. Mckinsey and Company (n.d.), By 2025, Internet of Things applications could have $11 trillion impact. https://www.mckinsey.com/industries/semiconductors/our-insights/whats-new-with-the-internet-of-things
8. Krishnan, P., Jain, K., Buyya, R., Vijayakumar, P., Nayyar, A., Bilal, M., & Song, H. (2021). MUD-based behavioral profiling security framework for software-defined IoT networks. *IEEE Internet of Things Journal*.
9. Kobusińska, A., Leung, C., Hsu, C. H., Raghavendra, S., & Chang, V. (2018). Emerging trends, issues and challenges in Internet of Things, big data and cloud computing. https://phoenixnap.com/blog/edge-computing-vs-cloud-computing
10. https://www.wwt.com/article/show-me-the-money-drive-new-revenue-streams-with-edge-computing
11. Sahu, S. K., Mohapatra, D. P., Rout, J. K., Sahoo, K. S., & Luhach, A. K. (2021). An ensemble-based scalable approach for intrusion detection using big data framework. *Big Data, 9*(4), 303–321.
12. Rout, S., Sahoo, K. S., Patra, S. S., Sahoo, B., & Puthal, D. (2021). Energy efficiency in software defined networking: A survey. *SN Computer Science, 2*(4), 1–15.
13. Nithya, S., Sangeetha, M., Prethi, K. A., Sahoo, K. S., Panda, S. K., & Gandomi, A. H. (2020). SDCF: A software-defined cyber foraging framework for cloudlet environment. *IEEE Transactions on Network and Service Management, 17*(4), 2423–2435.
14. Sahoo, S., Sahoo, K. S., Sahoo, B., & Gandomi, A. H. (2020, December). An Auction based Edge Resource Allocation Mechanism for IoT-enabled Smart Cities. In *2020 IEEE Symposium Series on Computational Intelligence (SSCI)* (pp. 1280–1286). IEEE.
15. Mishra, S. K., Mishra, S., Alsayat, A., Jhanjhi, N. Z., Humayun, M., Sahoo, K. S., & Luhach, A. K. (2020). Energy-aware task allocation for multi-cloud networks. *IEEE Access, 8*, 178825–178834.
16. Sahoo, K. S., Tiwary, M., Luhach, A. K., Nayyar, A., Choo, K. K. R., & Bilal, M. (2021). Demand-supply based economic model for resource provisioning in industrial IoT traffic. *IEEE Internet of Things Journal*.

17. Krishnamurthi, R., Nayyar, A., & Solanki, A. (2019). Innovation opportunities through Internet of Things (IoT) for smart cities. In *Green and Smart Technologies for Smart Cities* (pp. 261–292). CRC Press.

18. Tiwary, M., Puthal, D., Sahoo, K. S., Sahoo, B., & Yang, L. T. (2018). Response time optimization for cloudlets in mobile edge computing. *Journal of Parallel and Distributed Computing*, *119*, 81–91.

19. Neha, B., Panda, S. K., Sahu, P. K., Sahoo, K. S., & Gandomi, A. H. (2022). A systematic review on osmotic computing. *ACM Transactions on Internet of Things*, *3*(2), 1–30.

20. Bhoi, S. K., Panda, S. K., Jena, K. K., Sahoo, K. S., Jhanjhi, N. Z., Masud, M., & Aljahdali, S. (2022). IoT-EMS: An Internet of Things based environment monitoring system in volunteer computing environment. *Intelligent Automation and Soft Computing*, *32*(3), 1493–1507.

21. Maity, P., Saxena, S., Srivastava, S., Sahoo, K. S., Pradhan, A. K., & Kumar, N. (2021). An effective probabilistic technique for DDoS detection in OpenFlow controller. *IEEE Systems Journal*.

22. Bhoi, A., Nayak, R. P., Bhoi, S. K., Sethi, S., Panda, S. K., Sahoo, K. S., & Nayyar, A. (2021). IoT-IIRS: Internet of Things based intelligent-irrigation recommendation system using machine learning approach for efficient water usage. *PeerJ Computer Science*, *7*, e578.

3

SDN-Aided Edge Computing-Enabled AI for IoT and Smart Cities

Rashandeep Singh, Sunil Kr. Singh and Sudhakar Kumar
Chandigarh College of Engineering and Technology, Chandigarh, India

Shabeg Singh Gill
IIIT, New Delhi, India

CONTENTS

DOI: 10.1201/9781003213871-3

3.1 Introduction

Over the past few years, the volume of data being produced by internet of things (IoT) devices has seen explosive growth. A growing number of IoT devices share information through the sensors attached to them to provide optimum solutions requiring less time and less human intervention. The applications which have emerged to serve the concept of IoT include smart cities, medical and healthcare, education, agriculture, smart homes and the automotive industry [1–3]. The data produced through IoT devices is uploaded to the cloud for handling computing tasks [4]. High network latency makes the transmitting of data to the cloud through conventional cloud computing paradigms impractical : a major bottleneck of traditional IoT services is the network bandwidth [5].

An edge computing (EC) paradigm recently proposed to solve this bottleneck has received considerable attention. In the EC paradigm, all computing tasks happen at the edge of the network [6, 7]. It offers many benefits over cloud computing such as dramatically reduced latency during information transmission, protecting privacy of users, reducing the energy intake of information centers, and reducing the load of network bandwidth. The data produced by many smart devices is no longer communicated to the central cloud platform but is transmitted, stored and computed on the edge nodes to reduce latency time [8].

Communicating every bit of the enormous amount of data produced from numerous IoT devices is a challenging real-time task that can be met by using software-defined network (SDN) infrastructures. This technology, which was devised by a research team at Stanford University, has proved very useful in communications and networking domains.

In the traditional network, each device has a control plane associated with it along with the data forwarding plane. SDN, however, acts as a centralized control plane and has every update of every device present in the network [10]. SDN network infrastructures provide many benefits such as dynamic data flow, fast processing, easy monitoring, as well as manageable and cost-effective networks.

In the hierarchy of cloud servers and edge devices, the flow of data from cloud servers to edge nodes is called reverse offloading, as opposed to forward offloading where data flows from edge to cloud servers [11]. This decreases the latency of the network and computational cost, further improving overall quality of service (QoS) and quality of experience (QoE) of IoT devices.

For big data coming from IoT devices, technologies such as artificial intelligence (AI) and machine learning (ML) are utilized to analyse valuable sources of information [12]. The purpose is to enhance decision making, to automate and expand productivity, and to initiate new business lines [13]. Deep learning is used to protect privacy; the pre-processed data are preserved as data semantics and the meaning of intermediate data varies from that of the original data. The combination of all these technologies contributes to improved flexibility and safety.

IoT devices produce real-time information which can be used to provide better automation, as well as improved efficiency and productivity. However, a large amount of data requires high computations. The combination of SDN-aided edge computing and AI refine performance [14], with reduced energy requirements. The increasing involvement of these technologies is helping to revolutionize cities, making them sustainable or green cities [15]. The key objectives of this chapter are as follows:

- To study, how these technologies have helped to provide the best solutions for IoT devices, as well as the various challenges and opportunities that lie ahead.

- Illustrate how these modern technologies will enhance living standards in cities and can even make rural areas into smart cities.

- Highlight how to make rural areas smart to ensure equal opportunities in both urban and rural areas and that no individual city should be under pressure of resources.

- To simulate more research studies to provide equal and sustainable environments for facilitating industrial growth to promote globalization.

Organization of the Chapter

Section 3.2 provides insights into IoT and its various trends. Section 3.3 discusses SDN Section 3.4 provides details of the rise of AI. Section 3.5 describes edge, fog and cloud computing in detail and discusses the difference between them. Section 3.6 describes SDN-aided edge-aided AI for IoT. Section 3.7 discusses recent trends in edge computing-based IoT applications. The architecture and security of edge computing is discussed in Section 3.8. Section 3.9 deals with challenges and opportunities. In Section 3.10, there is a case study on SDN-aided EC, and Section 3.11 concludes the chapter with future scope.

3.2 Internet of Things (IoT)

The internet can be described as a communication network of computers and servers connected globally by routers and switches that uses an internet protocol suite (TCP/IP) to connect between networks and devices. In its early days, the internet sought to connect humans to communicate and share data, and then progressed to connecting things with the internet [16]. Things include smart devices which are embedded with sensors and have the capability to communicate [17]. These devices include smartphones, ovens, cars, air conditioners and other smart devices as shown in Figure 3.1.

IoT influences the way we react and behave in our daily lives [18]. With IoT, all these devices can exchange vast amounts of data over a common platform and with a common language to communicate. Every device directs the data securely to the IoT platform where collected data are integrated to perform added analytics [19]. The technology installed helps the things to interact with internal and external surroundings and make decisions. Depending on the requirements, valuable information is mined. The results are then shared with other devices.

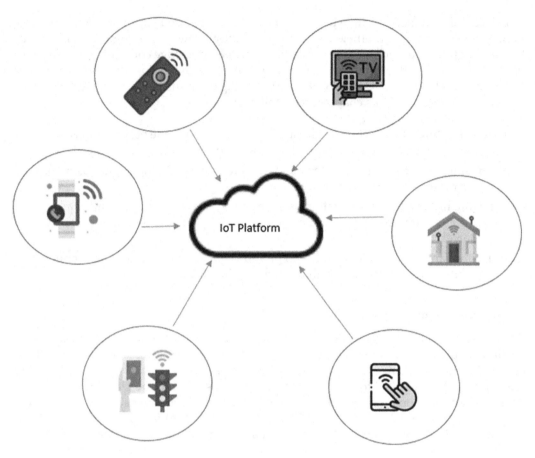

FIGURE 3.1
Common IoT platform for all devices.

IoT is seeking to revolutionize the world [20]. Day by day, more devices are being added to the giant network of IoT devices [21]. According to a report by Statistica [22], by 2025 there will be approximately 75.44 billion devices across the globe (Figure 3.2). These devices make decisions on their own without any human intervention. Among the results are improved user experience, automation and increased efficiency.

So, IoT is a multidisciplinary paradigm in which countless devices that surround us are linked to the internet with intelligent behaviour, and communicate through the internet to provide new services and improve efficiency [23]. This allows devices to learn from the experience of other devices as humans do.

3.2.1 Background to IoT

It was not until 1999 that the internet of things was publicly given a name [24]. Kevin Ashton, co-founder of MIT's Auto-ID Lab, coined the term in a presentation to Proctor and Gamble on production process optimization. It was his observation that the optimization of devices depends on the speed of data processing and transmission. Using tags, computers will be able to handle, trail and inventory all the devices without human intervention. Tags can be applied to things using barcodes, digital watermarks, and QR codes,

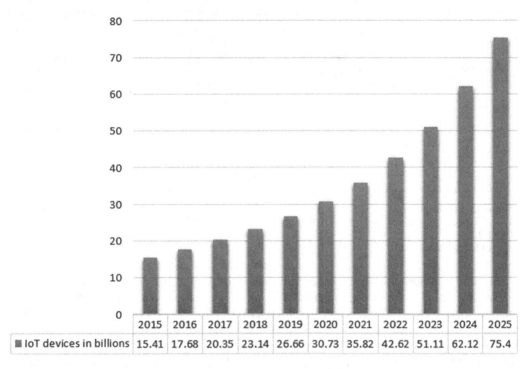

IoT devices in billions

	2015	2016	2017	2018	2019	2020	2021	2022	2023	2024	2025
▪ IoT devices in billions	15.41	17.68	20.35	23.14	26.66	30.73	35.82	42.62	51.11	62.12	75.4

FIGURE 3.2
IoT devices in billions from 2015 to 2025.

among other technologies. Ashton pioneered radio frequency identification (RFID), which is used in barcode detectors for supply-chain management, enabling data to be transmitted directly between devices without human intervention, and thus accelerating the process.

After almost a decade, the term 'IoT' began to be utilized in daily life. Along with AI, IoT has become an important concept in the progress of information technology (IT) [25]. It has developed into a machine using multiple technologies, starting from the internet to wireless communication and from micro-electromechanical systems (MEMS) to embedded systems. A wide variety of fields, including wireless sensor networks, GPS systems, and smart devices, all support IoT. Without much realizing it is observed that, IoT applications are becoming important aspects of our everyday lives.

3.2.2 Working Model of IoT

The general stages of IoT applications include collection, processing, storage and transmission (Figure 3.3). Every application requires the first and last stages, while some applications do not require the processing and storage stages.

An IoT device must have an internet connection, embedded sensors that will sense the environment and transmit valuable data to the IoT platform, functional software, in-built technologies which support network connections, and actuators such as a remote dashboard to monitor the output and display it in convenient form.

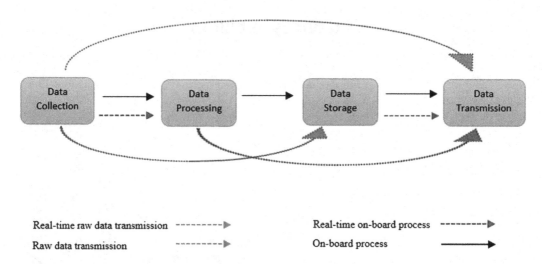

Real-time raw data transmission -------▷   Real-time on-board process -------▷

Raw data transmission   -------▷   On-board process   ⟶

FIGURE 3.3
Stages of IoT applications.

All these functionalities are combined together to form an IoT device. Smartphones are an example of IoT devices which permits a user to see traffic on roads through maps, find parking spaces, etc. It also has features such as Wi-Fi and Bluetooth which allows smartphones to connect with other devices such as smart televisions, laptops, and smart watches. Fit bands allows the user to see heartrate, calorie count, steps walked, etc. If there is a health problem, fitness devices send data to the hospital, and by the time the patient reaches hospital, the health report has already been seen by doctors and treatment can start quickly.

3.2.3 IoT Architecture

An architecture for IoT devices includes four layers (Figure 3.4). The bottom layer includes all smart devices such as phones, watches, remotes, etc,. with capability to sense, compute and connect other devices. The second layer, IoT gateway or aggregation layer aggregates the data from numerous sensors. This layer along with the base layer forms the definition engine.

The third layer, the processing engine or event processing engine, is based on the cloud. The data obtained from the sensors is processed using a variety of algorithms. The fourth layer, or API management layer, provides the interface with third-party applications and infrastructure.

The architecture is supported by device and identity managers that serve as valuable tools for ensuring safety and security.

3.2.4 IoT Technology Stack

The IoT technology stack can be broken down into four layers (Figure 3.5).

The IoT technology stack includes four layers: device hardware, device software, communication, and platform.

Device hardware: Devices act as an interface between the digital and real world and are actually the things within the internet of things (IoT). Devices comprise of microcontrollers, sensors to sense the data, and actuators to perform specific actions [26, 27] using embedded systems [28].

Device software: The devices have functional software and in-built technologies for implementing communication with the cloud and performing real-time data analysis within the IoT network. This functional software makes the devices 'smart'.

Communication: Data gathered by the devices is directed to a common platform to communicate with each other. Communication is an important part of building an IoT network. Various communication protocols are shown in Figure 3.6 [29].

Platform: Using IoT, all devices are able to dump their data and communicate using a common language. Platforms can be either on the premises or cloud based.

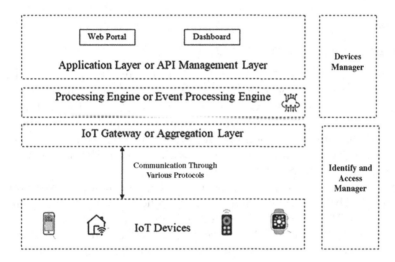

FIGURE 3.4
Architecture of IoT.

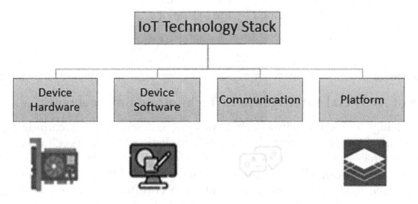

FIGURE 3.5
IoT technical stack.

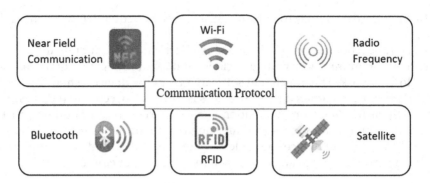

FIGURE 3.6
Communication protocols.

3.3 Software-Defined Network (SDN)

There is no predefined path for transmitting data from source to destination. Traditionally, it was decided by the router or the network devices. All routers take decisions independently of each other. A router is divided into two parts based on its functionality: control plane and data forwarding plane. The control plane is the part where all the routing protocols work and also prepares a routing table. The data forwarding plane takes data from outside and asks the control plane what to do with the data. The data is then sent to the forwarding plane of one of the next routers.

In contrast, SDN takes the control plane out of the network devices and provides a centralized controller. The forwarding plane of every device will ask the SDN controller what to do with the data. Therefore, every device in the network is updated using SDN.

There are three parts to the SDN architecture: infrastructure layer, control layer, and application layer (Figure 3.7). The network devices consisting of data planes are part of the infrastructure layer. The SDN controller is present at the control layer and is called the brain of this architecture. Communication between forwarding and control layers and also among network devices is done using OpenFlow protocol [30]. The application layer consists of various pieces of software to manage security, balance load, and manage traffic, etc. The software gives instructions to the controller, resulting in improved transmission speed and monitoring of the data generated by the IoT device.

3.4 Rise of Artificial Intelligence in Technology Enlargement

Artificial intelligence (AI) structures help to process issues such as the learning capabilities of computer machines, handling of languages, and distinguishing speech in an exceptionally solid and precise way. [9]. These tasks are achieved through various machine learning (ML) models or algorithms [31, 32] and deep learning (DL). ML and DL are subsets of artificial intelligence. Essentially, these models refer to algorithms that are made and prepared in the cloud or at a server center and are then conveyed to the edge devices. ML models are maintained and improved through continuous retraining and deployment. The improved models are then deployed at the edge devices.

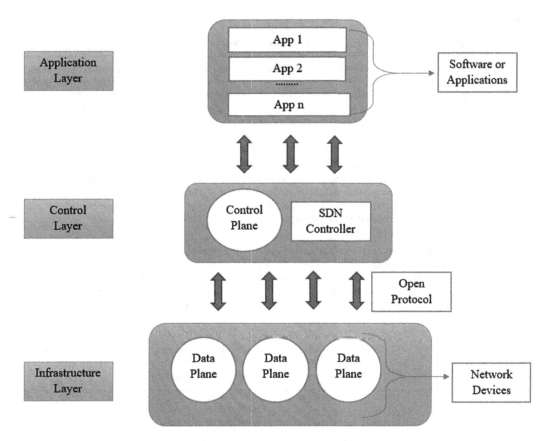

FIGURE 3.7
SDN architecture.

AI has been adopted in numerous arenas and enterprises, providing a multiplicity of innovative applications ranging from computer vision to speech recognition that are widely used in large-scale enterprises and local government operations [33]. It also provides services that are essentially determined by big data, machine learning calculations, high-performance computing, and storing amenities in the cloud. The value of the AI sector will grow to US$190.61 billion by 2025 [34]. The use of AI in technology over the last five years and predictions for the next five years are shown in detail in Table 3.1.

From one perspective, AI can provide edge computing technologies and techniques that can be utilized by edge computing, and edge computing can unlock its latent potential and be versatile with AI; while from another perspective, edge computing can provide AI situations and stages, and AI can increase its relevance with edge computing [41].

3.4.1 Technologies and Techniques Provided by Artificial Intelligence for Edge Computing

As a general rule, Edge Computing is a model where software-defined networks decentralize data and offer types of assistance with vigor and versatility. A number of layers in edge computing can suffer from resource allocation problems, including CPU cycle frequency, data transmission, radio frequency, and others. Therefore, it has remarkable requests for powerful optimization tools in order to improve its effectiveness. AI advances

TABLE 3.1

Use of AI Over Last 5 Years and Prediction for the Next 5 Years

	Use of AI Over Last Five Years	Prediction for Use of AI in Next Five Years
AI in Marketing	Industries are using AI to enhance their marketing applications such as prediction of customer behaviour [35], virtual assistants [36], etc.	In the coming years, industries will continue to use AI to be more effective and in order to make immediate decisions, understand the goals of consumers, develop campaigns for advertising, etc.
AI in Healthcare	AI has augmented a large number of applications in healthcare such as smart watches or fitbands to keep a check on blood pressure level, sugar level, etc., patient risk analysis [37], breast cancer detection [38], etc.	In future, hospitals will start using AI-driven applications in daily processes to real-time data analysis, helping doctors to make accurate decisions for patient monitoring.
AI in Automotive Industry	AI has been used in the automotive industry [39] over the past few years in various applications such as gesture and voice recognition systems.	In future, AI will be used as a major asset for self-driving cars where it will be used to process huge volumes of data to analyse situations on the road and respond accordingly. Moreover, AI will be used to augment road and vehicle security.
AI in Security	Over the past few years, with increase in use of AI, cyberattacks are also growing [40].	Security solutions integrated with AI will be used to protect data and strengthen network security.

are able to deal with this task. In principle, AI models determine asymptotically optimal solutions iteratively with SGD strategies by selecting unconstrained optimization problems from genuine situations. Either deep learning strategies or statistical learning methods are capable of offering assistance and guidance for the edge.

3.4.2 Edge Computing Gives AI Scenarios and Platforms to Work With

The IoE has become a reality because of the proliferation of IoT devices. Data is produced in large volumes from IoT devices such as driverless cars, smart cities, smart homes. Data processing in real time in public security can make AI a practical tool. AI applications that require low processing power and good communication quality can also be moved from the cloud to the edge.

3.5 AI-Enabled IoT Computing

IoT and AI, especially AI-enabled IoT, are critical to understanding ubiquitous intelligence [42]. The vast amount of data coming from IoT devices – which is known as big data – is a valuable source of information. By utilizing AI and ML, data is analysed continuously to promote better decision making and automation, as well as in order to expand productivity and initiate new business lines [43]. Different types of computing are described in the following sections.

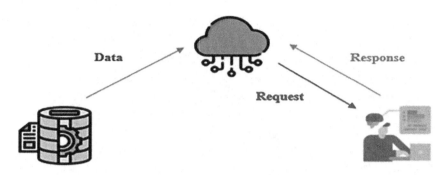

FIGURE 3.8
Cloud computing paradigm.

3.5.1 Edge Computing

Recent years have seen edge computing becoming increasingly popular. Because there is an increasing trend to deliver data at the edge of the organization, it is much more productive to handle the information at the edge as well [44]. A previous study by cloudlet [45] and fog computing [46] found that cloud computing might not be a good choice for processing data created at the edge of the network. But the cloud's processing power dominates capacity of the things at the edge; pushing all the computing tasks onto the cloud has proven to be an effective method to handle information. An orthodox cloud computing paradigm is shown in Figure 3.8. Producers of data produce raw information, which makes its way to the cloud as indicated by the blue line. Data consumers request information from the cloud using the red line, and the cloud provides the resulting information via the green line.

However, because of the overwhelming quantity of data at the edge, the network bandwidth has, in comparison to the rapidly growing data processing speed, come to a halt [45]. As a result, the demand for data speed and transit becomes a bottleneck for cloud-based computing as the volume of data produced at the edge increases. An airplane like the Boeing 787, for instance, creates around 5 gigabytes of data every second [46], yet the bandwidth between the airplane and either a satellite or ground-based base station is not sufficient to transmit this data. Think about another example of an independent vehicle. Gigabytes of data will be generated by the vehicle on a consistent basis, and the vehicle must process the data all the time to come up with the right decisions. For example, a vehicle's camera captures an enormous amount of video data, which the system should process in order to continue to provide good driving decisions. Data sent from the vehicle to the cloud would require an excessive response time. Moreover, having countless autonomous vehicles in one region would further strain network capacity and reliability [47]. The reaction time would be unreasonable if the cloud were to be utilized for processing each and every piece of data. Moreover, the data transfer capacity and dependability of the platform would be tested in order to support a large fleet of vehicles in one region. It would be prudent to handle the data at the edge in this case, so as to achieve a quicker response time, more effective real-time processing and a reduced network pressure.

Large-scale data analytics have become possible due to the user-driven nature of many cloud services. On the other hand, centralized cloud servers simply increase the frequency of communication between users and their devices, like wearables, tablets, mobile phones, etc. (called edge devices) and geographically distant cloud data centers, servers, databases,

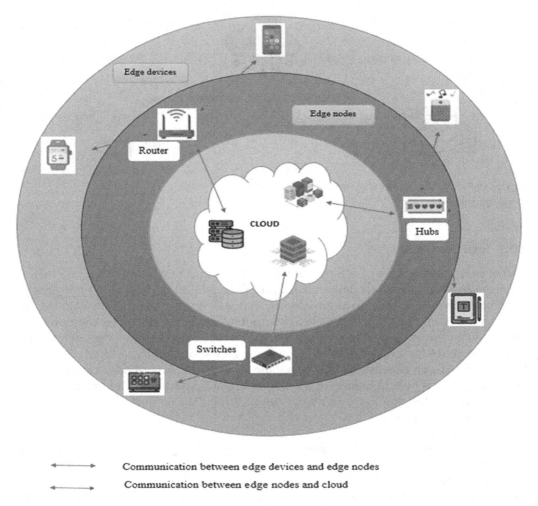

Communication between edge devices and edge nodes

Communication between edge nodes and cloud

FIGURE 3.9
Relation of edge devices, nodes and cloud.

etc. Applications that rely on real-time reactions are affected by this limitation. Therefore, there has been increased interest in exploring "beyond the cloud" to the network's edge (Figure 3.9), a concept known as edge computing [48, 49]. Additionally, the calculations are performed on nodes such as switches and routers which helps with traffic coordination.

So, IoT services are also dependent on edge computing [50–52]. Inefficient cloud computing is occurring as a result of data transfer with limited network capacity, resulting in increased inefficiency in processing and analysing the huge volume of data gathered from IoT devices [53, 54]. Using pre-processing techniques significantly reduces the amount of data transferred since edge computing decentralizes computing to devices near the edge instead of a centralized cloud. The edge computing platform aids the IoT in resolving several crucial difficulties and refining performance. As demonstrated in Figure 3.10, edge and IoT have several traits in common.

The learning network in a DL model usually contains multiple layers. Each network layer can swiftly scale down the intermediary data size till sufficient features are perceived. DL is therefore well suited for edge computing due to its ability to offload learning layers

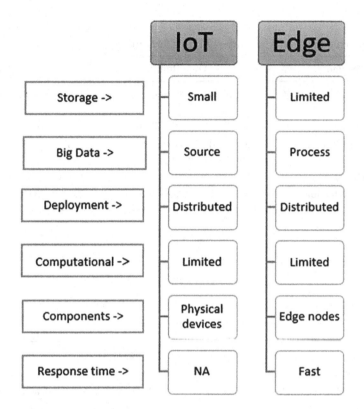

FIGURE 3.10
Characteristics of IoT and edge computing.

at the edge and to reduce intermediate data so that it can be transported to centralized cloud servers [55].

Additionally, DL in edge computing protects your privacy while you transfer intermediate data. The user privacy is maintained in typical big-data systems because the preprocessing is preserved as data semantics. Often, intermediate data in DL has a different meaning from the original data. For instance, the intermediate layer makes it very difficult to understand the original data with features derived from a convolutional neural network (CNN) [56].

As summarized in Table 3.2, edge computing presents a number of motivations, opportunities, and challenges.

3.5.2 Fog Computing

SISCO defines fog computing as "the use of highly virtualized platforms by end devices to access cloud computing data centers, which are typically located at the edge of the network"[57]. A fog computing platform provides networking, computation, and storage services through highly virtualized platforms that are connected to traditional cloud computing environments [58, 59]. This process is often referred to as execution of edge computing [60–62]. Fog computing's purpose is to bring networking capabilities, dispersed processing, storage, and control closer to the user [63]. Table 3.3 compares edge computing with fog computing.

TABLE 3.2

Motivation, Opportunities, and Challenges of Edge Computing

Motivation	Opportunities	Challenges
Overcoming resource constraint of front-end devices	Lightweight libraries	Discovery of edge nodes
Decentralized computing	More efficient algorithms	Security of edge nodes
Dealing with network traffic	Standards, benchmarking and marketplace	General-purpose computing on edge nodes
Low latency computing	Languages and frameworks	Excellent quality of service (QoS)
Smart computation techniques	Micro operating systems	Partitioning and offloading tasks

TABLE 3.3

Edge v/s Fog Computing

Edge Computing	Fog Computing
It is less reliable	It is highly reliable
It is very low latency	It has low latency
It is generally accessed through LAN or WLAN	LAN is used as a network accessory
It has high response time	It has slightly low response time compared to EC
Network is largely distributed	Its network is moderately distributed
There is no possibility of standardization	It can be standardized
Server scalability of it is high and dispersed	It has low and dispersed server scalability
It has less storage as it is much closer to location of data generation	It also has less storage for the same reason as edge computing

3.5.3 Cloud Computing

NIST [64] defines cloud computing as "a model for enabling ubiquitous [65, 66], convenient and its evaluation, on-demand network access to digital resources (e.g., networks, servers, storage, and applications) that can be quickly provisioned and released with minimum management effort or service provider input". Cloud computing reference design is proposed in [67]. An introduction to cloud computing and an explanation of how it works can be found in [68, 69]. In a cloud computing environment, there are two layers (the end device and the cloud data center), while in edge computing, there are three, four, or five layers. The edge paradigm facilitates both networking and storage close to users, as well as processing at the data center. With cloud computing, all data processing takes place in a central data center, and end devices have to wait for the data center to return the processed results. Table 3.4 summarizes the differences between edge and cloud computing in regard to their specialized boundaries and general attributes [70–72].

TABLE 3.4

Edge v/s Cloud Computing

Edge Computing	Cloud Computing
It has a local service scope	It has global service scope
It generally has low latency	It has high latency
The data is generated by sensors, humans, or devices	Most of the data is generated by humans
It follows a distributed network paradigm	It has centralized geo-distribution
Mobility is supported	Mobility support is limited
It consists of large nodes including router, switches, etc.	It has very few server nodes including data centers, databases, etc.
It has awareness of location	It does not have location awareness
The target users are mobile users	It covers general internet users
It consists of low storage	It has high storage
There is no possibility of standardization	Standardization is possible
The distance between server and client is single hops	The distance between server and client is multiple hops

3.6 SDN-Aided Edge-Enabled AI for IoT

As a result of the complex, diverse, and dense environment of interconnected IoT devices with both edge and cloud systems, management functions must be automated [73]. Several studies have shown that SDN can effectively address many limitations of conventional IoT, including system control, automation, and security concerns [74]. The use of SDN-supported edge-aided AI for IoT facilitates reliable and secure processes and has the benefit of scalability [75]. The general architecture of the SDN-aided EC environment is shown in Figure 3.11.

Some of the benefits include:

- Task offloading: SDN-aided edge can facilitate in task offloading, with alternative devices executing the task on behalf of a local device [76].
- Resource allocation: SDN located between edge and cloud resolves the issue of resource planning and allocation by providing dynamic routing routes to avoid network bottleneck [77, 78].
- Load balancing: SDN helps to distribute the path computation load which can reduce communication overheads and end-to-end delays [79, 80].
- Scalability: SDN helps make the network more scalable and flexible and allows new IoT devices to be the part of that system.

With the number of IoT devices increasing daily, security threats are becoming more sophisticated, harder to control, and more difficult to detect. The continuous flow of data packets from the large number of IoT devices can lead to link disruption, which further degrades QoE and QoS for various IoT edge applications. Various AI models or algorithms can be applied in the control layer to predict failure [81]. These will further assist the SDN controller to choose a consistent communication path or link for IoT devices to ensure better QoE, QoS, and end-to-end latency.

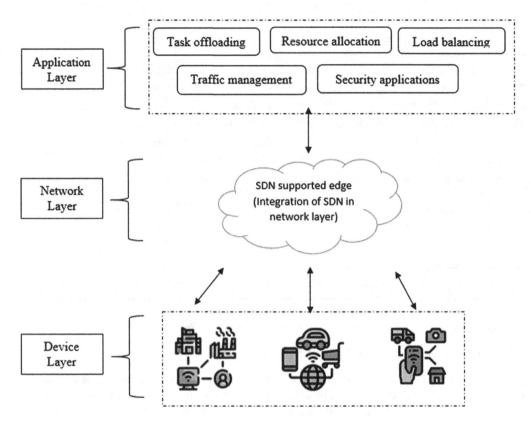

FIGURE 3.11
Architecture of SDN-aided EC environment.

SDN in the hierarchy of edge nodes and cloud data centers helps in the movement of information, and the combination of edge and SDN can lead to flexibility, mobility, and management of the network [82]. Therefore, SDN-aided edge computing for IoT provides a seamless and efficient flow of data from edge nodes to cloud computing server or vice versa. Combining all these technologies helps solve various problems and complexities inherent to IoT networks.

3.7 Recent Trends in Edge Computing-Based IoT Applications

Many applications have emerged to serve the concept of SDN-aided edge computing for IoT [83].

3.7.1 Smart Cities

Smart cities are designed to improve living conditions, merging technology, society and government to enable a smart environment, smart mobility, smart living, and smart governance [84, 85]. As IoT is adopted across cities, city infrastructure will become more efficient and citizens will be better served [86]. Smart cities provide significant innovations covering a wide variety of applications including smart lighting, smart parking, waste management, water management etc.

- *Smart lighting:* Sensors collect data about pedestrian and traffic conditions and provide optimum lighting.
- *Smart traffic management:* Traffic problems are increasing due to rapid urbanization, with the number of vehicles proliferating every day. With the assistance of dynamic traffic modeling and real-time data from traffic sensors and cameras, managing traffic congestion will be much easier and smart traffic signals will help better traffic management.
- *Smart parking:* Sensors send data to the edge and further to the cloud. The data create a virtual parking map and users can use the application to view available parking spaces.
- *Waste management:* Dustbins have sensors which measure the threshold capacity and send information to the relevant department to collect the rubbish once it is overflowing. This information helps optimize garbage truck routes.
- *Water management:* Smart water monitoring devices helps in water conservation, smart irrigation, leakage management and water quality management. Potable water monitoring tools help to monitor the quality of tap water.

3.7.2 Business

The success of the business entirely depends on how you deal with situations that arise throughout the process. Maintenance, flawless production, and customer satisfaction are some of the pillars of business performance. IoT is required in business to yield the best results needed to grow and lead the competition [87, 88]. It helps to improve internal operations and serve customers more efficiently.

3.7.3 Medical and Health Care

IoT technology helps save time and hence save lives. This technology is widely used in the healthcare sector to track the status, diagnosis and treatment of patients remotely [89] in both urban and rural areas [90]. IoT technology is being embedded in health care devices such as fit bands that help to monitor and improve the medical conditions of the patient. All the data collected on the IoT platform from numerous medical devices can be used to analyse patients' conditions and find appropriate treatment [91].

3.7.4 Education

IoT has a great impact on education institutions [92]. IoT makes education more accessible in terms of ability, status, and geography [93]. It leads to enhancement in teaching, learning cooperation, and student engagement, filling the gaps in the education industry by saving time through a fingerprint attendance system which requires a sensor and microcontroller. It also improves the quality of education through advanced smart classrooms and digital boards. IoT has also improved children's safety through bus mobile apps and smart security cameras that keep parents updated about their child's whereabouts.

3.7.5 Agriculture

IoT devices have helped rural areas to come into the mainstream of development and match the opportunities of urban areas [94]. IoT helps farmers to monitor fields, where embedded sensors give a holistic view of performance, help schedule servicing and prevent yield-sapping breakdowns [95]. Sensors gather real-time data about weather, air quality, soil and hydration levels that helps farmers make decisions about planting and harvesting of crops [96]. Sensors are used to measure soil moisture content, and data collected will be sent to an IoT platform. If the moisture content of a particular sector of the field is greater than the threshold value, there is no need for water supply. But if the moisture content is less than the threshold, water will be supplied in that particular sector through the water pipelines dynamically without any human effort, saving time and labor costs, and helping with water management.

3.7.6 Smart Home

The idea of a smart environment depends on integration between IoT and SDN-aided edge computing. A home is called smart when it consists of smart devices that are controlled remotely and can be set to your own requirements [97]. Smart devices communicate with each other to provide comfort to humans with minimum human efforts, saving time and power [98]. Lot of processes run autonomously in smart homes, including smart air quality adjustment, air conditioning temperature adjustment, lighting control, voice-based control, etc. Security features such as security cameras, smart locks with iris-based authentication, and motion sensors that alert users when activity is detected are the main reasons attracting people to smart home technology. All these processes have to be executed automatically in parallel without human intervention [99].

3.7.7 Automotive Industry

Implementation of IoT applications in automotive industry results in

- Enhanced performance,
- Reduced costs, and
- Quality control.

For instance, driver safety is an important factor that can be ensured using IoT technology. Sensors embedded in the vehicle provide information about the current status of the vehicle, monitoring the functionality of the engine, brakes, and electrical system. Operating cost is reduced and safety is increased.

Advanced IoT technology solutions in the automotive industry include connected cars, vehicle-to-vehicle (V2V) applications, vehicle-to-infrastructure (V2I) applications, and vehicle-to-everything (V2X) communication applications. A connected car is a vehicle with internet access that allows devices which can be inside or outside the car to interact with each other. Devices include smart driving assistants, GPS navigation, key finder, smartphone mount, etc. Several smart apps like car control systems are making their way to cars. Conceptually, connected cars are based on the idea that the moving vehicles will communicate with control devices such as traffic lights to help control traffic.

3.8 Edge Computing: Architecture and Security

This section gives an outline of the architecture and tasks involved with EC. It also summarizes research with respect to security assessment and hazard relief procedures to guarantee that EC clients/administrators can accomplish their security requirements.

3.8.1 Architecture and Tasks

The term edge computing refers to a distributed computing architecture that basically involves the handling of information collected. There is a need to limit both data transfer capacity and response time in an IoT framework. Edge computing is necessary when latency requires improvement in order to avoid saturation of the network, just as when processing requirements are high in a centralized infrastructure [100]. Edge computing is a detailed version of fog computing, which uses edge devices to perform calculations, build up storage and communicate geographically, all of which are inputs and outputs of our current reality, which we refer to as transduction. The fog nodes decide whether to handle the information locally or to send it to the cloud [101].

Figure 3.12 illustrates communication from edge gadgets, such as cell phones, TVs, and so forth, to fog nodes, and later to cloud data centers. A fog node is responsible for connecting the cloud to edge devices. Devices at the edge serve as interfaces between individual or different things associating with them and the cloud [101].

Figure 3.13 summarizes the major tasks involved in EC. As described in [102], input, processing, and output are the three basic elements.

- *Data sources:* Data sources are endpoints which collect and record information about their surroundings or from customers.
- *AI:* In addition to data gathering and processing, the processing function (machine learning and deep learning models) identify practical perceptions, patterns and trends, produce individualized proposals, and further develop the performance dependent on AI or data analysis.
- *Human–machine interaction (HMI):* People can only act and make informed choices when results from earlier stages are implemented. Consequently, during this stage, bits of information appear in a clear way in the form of control panels, visualizations, alarms, etc., which enables connectivity between machines and people, creating a valuable feedback loop.

3.8.2 Privacy and Security

Organizations are responsible for ensuring the security and integrity of their IoT systems. A glossary of privacy-preserving management terms is provided below [103].

- *Pseudonymity:* The pseudonym is utilized as an ID to guarantee that an individual can use the source (for example pseudonym), uncovering the source's genuine identity. The client may, however, be responsible for utilization.
- *Unlinkability:* This ensures that a third party (such as an attacker) cannot determine whether or not two objects are connected.

- *Unobservability:* Providing individuals with the ability to use resources or services without third parties while being able to see the usage of those resources or services.

- *Anonymity:* Using a resource does not require the individual's identity to be revealed.

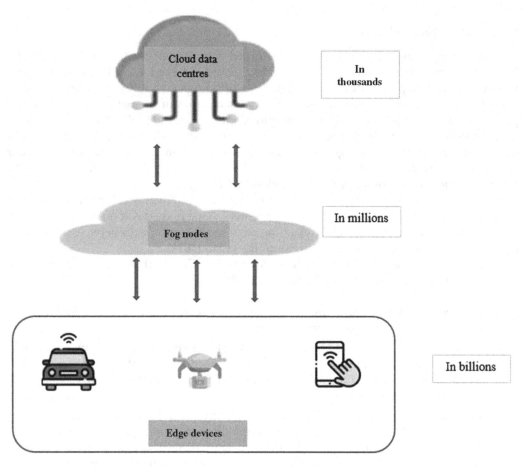

FIGURE 3.12
Typical edge computing architecture.

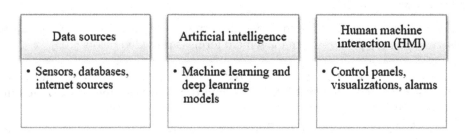

FIGURE 3.13
Major tasks of edge computing.

According to Zhang et al. [104], the evaluation of system security contains the following crucial components:

- *Privacy:* Ensuring that access to individual data in edge computing is only granted to the information owner and that individual. As soon as the information is moved and furthermore gathered or handled in organization systems at the edge or the center, access by unapproved parties is prevented.

- *Integrity:* Making sure that the information is legitimately and consistently transmitted to the authorized individual without unapproved changes. Security of people can be affected because of the absence of respectability measures.

- *Accessibility:* Ensuring that the licensed party knows how to provide edge administrations in any region based on what people need. Also, cipher text content stored on nodes at the edge or in the cloud can be handled under a variety of viable scenarios.

- *Access manage and verification:* In the access manage strategy, access manage mimics a central point of connection for all protection and security requests. A verification process ensures that a person's identity is confirmed.

3.8.3 Measures and Risk Reduction

A risk management organization should also monitor and describe IoT foundations' hazards using the methods listed below [105, 106]:

- *Solid Password Policy:* Make sure that people follow the ideal security secret key approach. A password should not be a thesaurus word, but rather a mix of lowercase and capital letters combined with a mixture of special characters. It is possible to create solid passwords using arbitrarily generated secret keys.

- *Encryption:* Organizations need to implement best-in-class ciphers to encrypt inbound and outbound communications, and in case of data loss, a disaster recovery back-up strategy should be developed.

- *Two-factor authentication* (2FA): Upon entering their username and password, people are prompted to authenticate their identities using 2FA. In addition, it will force another level of verification dependent on ATM PINs, secret keys, voiceprints, etc.

2FA can be further categorized as follows:

- *SMS messaging and voice-based:* For a secure login, the code can be obtained from SMS; and the numbers can be perused through a computerized voice call.

- *Hardware tokens:* Generates and displays a one-time secret key (OTP) for every exchange.

- *Software application tokens:* Rather than a conventional equipment token, a cell phone token uses a secure software application that can be downloaded to the client's cell phone.

- *Pop-up message:* Client's device receives a 'push' message as a second-factor confirmation to confirm the client's identity.

3.9 Challenges and Opportunities

The large number of sensors installed in IoT gadgets are undoubtedly impacting communications, guaranteeing a consistent association between people and gadgets. A number of issues arise from the increasing prevalence of sensors and the amount of information disposed of/created by them [107].

- *Client protection:* Client security in this day and age includes any data that might possibly reveal a client's character, conduct, or location. The objective of protecting a client's private data can counter the more extensive sending of IoT-empowered gadgets. A reliable framework is therefore one that is designed to gather and handle a lot of information without disclosing a client's private data.

- *Optimization metrics:* There are several layers with different calculation capacities in EC. Deciding which layer is to manage the complex responsibilities or the quantity of relegated tasks is a difficult task. Nevertheless, there are four considerations in determining an ideal allocation of responsibilities:
  a) Latency that is expected for system administration and calculation,
  b) Power utilization,
  c) Cost of development and maintenance, and
  d) Transmission capacity

- *Public openness of edge nodes:* When an edge gadget (for instance, a router, base station, switch, etc.) is expected to be utilized for free, many moves should be managed. A public/privately owned business needs to identify the dangers associated with their own gadgets deprived of compromising the ideal reason for the gadget (e.g., a switch) to be utilized as an edge hub. Multi-occupancy of edge nodes is just practical with present-day advances that prioritize security. Maintenance expenses, information areas, and responsibility for setting up proper value models making edge nodes promptly accessible, are also concerns.

To deal with the enormous quantity of information gathered by edge devices and sent to cloud data centers, pre-execution of information mining in EC can be performed. Extracting inadequate and dubious information is a major issue that requires advanced information mining calculations [108]. Energy can be saved using eco-accommodating assets like sun and wind. Environmentally friendly power assets include geothermal, biomass, and wind energy. In order to enable cost-efficient operation for better service quality, resource management algorithms should be adaptive in nature and perform management according to the nature of the application. A software/hardware resilient smart grid must use EC infrastructure to handle system failures.

Most networks depend on wireless network architectures that face a number of challenges to provide QoS and QoE. SDN helps to deliver a unified design to complex network architecture. The load-balancing feature of SDN facilitates optimum usage of services offered by SDN infrastructure. The programmable utility of SDN allows automatic and fast setup of networks. Additionally, SDN makes it possible for network service providers to integrate network elements into applications.

3.10 Case Study of SDN-Aided Edge Computing

This case study examines a smart banking queue framework that reduces congestion among customers at a bank looking for service.

Imagine a framework that permits people to instantly observe an accessible service by just clicking on a smartphone. This framework will reduce long hours spent waiting in a bank [109], and with the introduction of SDN into the hierarchy, network management and seamless data flow between edge servers and cloud servers will be made more efficient. The smart banking queue framework is normally controlled through RFID, motion sensors, ultrasonic finder, and infrared detecting units (see Figure 3.14). Figure 3.15 shows the usual execution process.

Individuals can benefit from this system as it can limit customer flow at a bank, which may be useful for those who do not want to waste time. It can also help the bank to upgrade and improve its services. In future, more effective algorithms can be set up to maximize the utilization of assets, such as the availability of time for banking services. For example, a deep learning model can be prepared to allocate space in real time.

3.11 Conclusion and Future Scope

IoT has become one of today's trending technologies. The data produced from the large number of IoT devices is sent to an edge computing platform. EC provides a promising aspect to smart devices with instantaneous computing power and storage. Moreover, with so many IoT devices on a network, SDN has become a crucial control technology for them.

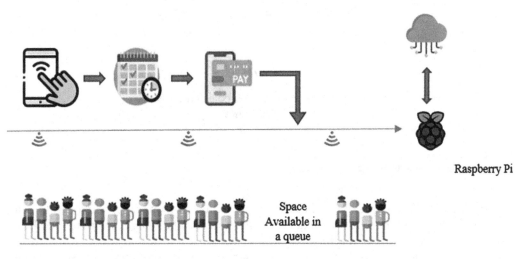

FIGURE 3.14
Example of smart banking queue system.

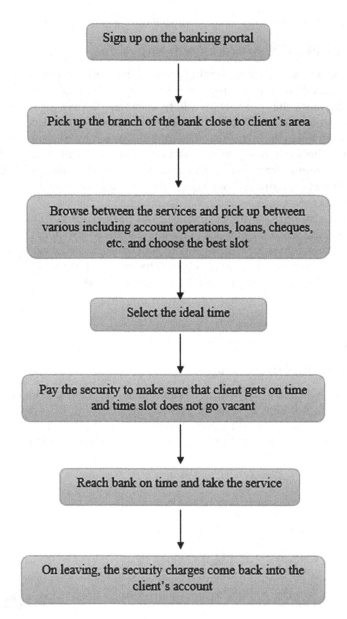

FIGURE 3.15
Flow of execution of smart banking queue system.

SDN diminishes the complexity at the edge node and helps with resource allocation, scalability, flexibility, and mobility of the network. Also, artificial intelligence has helped EC to enhance decision making and achieve data privacy.

This chapter has discussed architecture and security of edge computing as well as both challenges and opportunities in the field. SDN infrastructure and the benefits of its integration with EC have also been explored. SDN-aided edge computing-enabled AI for IoT and smart cities are also discussed. A case study on edge computing for smart banking queue is included to demonstrate the SDN-supported edge computing vision in depth. Although SDN-aided EC is still in its initial phase, we hope that this paper will encourage researchers and stimulate further research studies.

References

1. Ming, F. X., Habeeb, R. A. A., Md Nasaruddin, F. H. B., & Gani, A. B. 2019. Real-time carbon dioxide monitoring based on IOT & cloud technologies. In *Proceedings of the 2019 8th International Conference on Software and Computer Applications*, 517–521. Cairo, Egypt.
2. Moin, S., Karim, A., Safdar, Z., Safdar, K., Ahmed, E., & Imran, M. 2019. Securing IoTs in distributed blockchain: Analysis, requirements and open issues. *Future Generation Computer Systems*, 100, 325–343.
3. Chu, F., Yuan, S., & Peng, Z. 2006. Using Machine Learning Techniques to Identify Botnet Traffic. In *Encyclopedia of Structural Health Monitoring*. Wiley, Hoboken, NJ.
4. Singh, S. P., Nayyar, A., Kumar, R., & A. Sharma 2019. Fog computing: From architecture to edge computing and big data processing. *The Journal of Supercomputing*, 75(4), 2070–2105.
5. Kaur, A., Singh, P., & Nayyar, A. 2020. Fog Computing: Building a Road to IoT with Fog Analytics In *Fog Data Analytics for IoT Applications*, 59–78. Springer, Singapore.
6. Shi, W., & Dustdar, S. 2016. The promise of edge computing. *Computer*, 49(5), 78–81.
7. Xia, X., Chen, F., He, Q., Grundy, J., Abdelrazek, M., & Jin, H. 2020. Cost-effective app data distribution in edge computing. *IEEE Transactions on Parallel and Distributed Systems*, 32(1), 31–44.
8. Neha, B., Panda, S. K., Sahu, P. K., Sahoo, K. S., & Gandomi, A. H. 2022. A systematic review on osmotic computing. *ACM Transactions on Internet of Things*, 3(2), 1–30.
9. Kumar, V. A., Kumar, A., Batth, R. S., Rashid, M., Gupta, S. K., & Raghuraman, M. 2021. Efficient data transfer in edge envisioned environment using artificial intelligence based edge node algorithm. *Transactions on Emerging Telecommunications Technologies*, 32(6), e4110.
10. Aujla, G. S., Chaudhary, R., Kumar, N., Rodrigues, J. J., & Vinel, A. 2017. Data offloading in 5G-enabled software-defined vehicular networks: A stackelberg-game-based approach. *IEEE Communications Magazine*, 55(8), 100–108.
11. Aujla, G. S., Chaudhary, R., Kaur, K., Garg, S., Kumar, N., & Ranjan, R. 2019. SAFE: SDN-Assisted Framework for Edge–Cloud Interplay in Secure Healthcare Ecosystem. In *IEEE Transactions on Industrial Informatics*, 15(1), 469–480. doi: 10.1109/TII.2018.2866917.
12. Calo, S. B., Touna, M., Verma, D. C., & Cullen, A. 2017. Edge computing architecture for applying AI to IoT. In *2017 IEEE International Conference on Big Data (Big Data)*, 3012–3016. IEEE.
13. Ullah, F., Al-Turjman, F., & Nayyar, A. 2020. IoT-based green city architecture using secured and sustainable android services. *Environmental Technology & Innovation*, 20, 101091.
14. Hammoud, A., Sami, H., Mourad, A., Otrok, H., Mizouni, R., & Bentahar, J. 2020. AI, blockchain, and vehicular edge computing for smart and secure IoV: Challenges and directions. *IEEE Internet of Things Magazine*, 3(2), 68–73.
15. Chiesura, A. 2004. The role of urban parks for the sustainable city. *Landscape and urban planning*, 68(1), 129–138.
16. Solanki, A., & Nayyar, A. 2019. Green Internet of Things (G-IoT): ICT Technologies, Principles, Applications, Projects, and Challenges. In *Handbook of Research on Big Data and the IoT*, 379–405. IGI Global.
17. Krishnamurthi, R., Kumar, A., Gopinathan, D., Nayyar, A., & Qureshi, B. 2020. An overview of IoT sensor data processing, fusion, and analysis techniques. *Sensors*, 20(21), 6076.
18. Khan, M. A., & Salah, K. 2018. IoT security: Review, blockchain solutions, and open challenges. *Future Generation Computer Systems*, 82, 395–411.

19. Georgakopoulos, D., & Jayaraman, P. P. 2016. Internet of things: From internet scale sensing to smart services. *Computing*, 98(10), 1041–1058.
20. Singh, R., Angmo, R., Jha, V., Singh, P., Singh, V. P., & Aggarwal, N. 2021. Internet of Things (IoT) protocols, communication technologies, and services in industry. In *2021 3rd International Conference on Advances in Computing, Communication Control and Networking (ICAC3N)*, 1407–1413.
21. Alam, T. 2018. A reliable communication framework and its use in internet of things (IoT). *CSEIT1835111 | Received, 10*, 450–456.
22. Statista Research Department. 2016. Internet of Things – Number of connected devices worldwide 2015–2025. Statista. https://www.statista.com/statistics/471264/iot-number-of-connected-devices-worldwide/ (accessed: December 10, 2021).
23. Gubbi, J., Buyya, R., Marusic, S., & Palaniswami, M. 2013. Internet of Things (IoT): A vision, architectural elements, and future directions. *Future Generation Computer Systems*, 29(7), 1645–1660.
24. Suresh, P., Daniel, J. V., Parthasarathy, V., & Aswathy, R. H. 2014. A state of the art review on the Internet of Things (IoT) history, technology and fields of deployment. In *2014 International Conference on Science Engineering and Management Research (ICSEMR)*, 1–8. IEEE.
25. Ghosh, A., Chakraborty, D., & Law, A. 2018. Artificial intelligence in Internet of things. *CAAI Transactions on Intelligence Technology*, 3(4), 208–218.
26. Georgios, L., Kerstin, S., & Theofylaktos, A. 2019. Internet of things in the context of industry 4.0: An overview. *International Journal of Entrepreneurial Knowledge*, 7(1), 4–19.
27. Gupta M., & Singh S. 2019. The Internet of Things: An overview of the awareness, architecture & application. *International Journal of Latest Trends in Engineering and Technology*, 12(4), 19–24.
28. Singh, S. K., Bhatia, M. P. S., & Jindal, R. 2009. Architectural modeling for hardware and software in reconfigurable embedded system. *International Journal of Recent Trends in Engineering*, 1(1), 575.
29. Al-Sarawi, S., Anbar, M., Alieyan, K., & Alzubaidi, M. 2017. Internet of Things (IoT) communication protocols. In *2017 8th International Conference on Information Technology (ICIT)*, 685–690. IEEE.
30. Sinh, D., Le, L. V., Lin, B. S. P., & Tung, L. P. 2018. SDN/NFV—A new approach of deploying network infrastructure for IoT. In *2018 27th Wireless and Optical Communication Conference (WOCC)*, 1–5. IEEE.
31. Aggarwal, K., Singh, S. K., Chopra, M., & Kumar, S. 2022. Role of Social Media in the COVID-19 Pandemic: A Literature Review. In *Data Mining Approaches for Big Data and Sentiment Analysis in Social Media*, 91–115. IGI Global.
32. Chopra, M., Singh, S. K., Aggarwal, K., & Gupta, A. 2022. Predicting Catastrophic Events Using Machine Learning Models for Natural Language Processing. In *Data Mining Approaches for Big Data and Sentiment Analysis in Social Media*, 223–243. IGI Global.
33. Zou, Z., Jin, Y., Nevalainen, P., Huan, Y., Heikkonen, J., & Westerlund, T. 2019. Edge and fog computing enabled AI for IoT-an overview. In *2019 IEEE International Conference on Artificial Intelligence Circuits and Systems (AICAS)*, 51–56. IEEE.
34. Singh, S. 2019. The impact of artificial intelligence over the next five years. Forbes.
35. Gkikas, D. C., & Theodoridis, P. K. 2022. AI in Consumer Behavior. In *Advances in Artificial Intelligence-Based Technologies*, 147–176. Springer, Cham.
36. Garíca-Serrano, A. M., Martínez, P., & Hernández, J. Z. 2004. Using AI techniques to support advanced interaction capabilities in a virtual assistant for e-commerce. *Expert Systems with Applications*, 26(3), 413–426.
37. Pourhomayoun, M., & Shakibi, M. 2020. Predicting mortality risk in patients with COVID-19 using artificial intelligence to help medical decision-making. *MedRxiv*.
38. Voth, D. 2005. Using AI to detect breast cancer. *IEEE Intelligent Systems*, 20(1), 5–7.
39. Gusikhin, O., Rychtyckyj, N., & Filev, D. 2007. Intelligent systems in the automotive industry: Applications and trends. *Knowledge and Information Systems*, 12(2), 147–168.

40. Kaloudi, N., & Li, J. 2020. The ai-based cyber threat landscape: A survey. *ACM Computing Surveys (CSUR)*, 53(1), 1–34.

41. Deng, S., Zhao, H., Fang, W., Yin, J., Dustdar, S., & Zomaya, A. Y. 2020. Edge intelligence: The confluence of edge computing and artificial intelligence. *IEEE Internet of Things Journal*, 7(8), 7457–7469.

42. Zheng, L.-R., Tenhunen, H., & Zou, Z. 2018. *Smart Electronic Systems: Heterogeneous Integration of Silicon and Printed Electronics*. John Wiley & Sons.

43. Gupta A., Singh S., Chopra M. 2020. Impact of artificial intelligence and Internet of Things in modern times: A research paper. *Self-reliant India and Technology Teaching (Special Issue), Vigyan Garima Sindhu*, 113, 81–97.

44. Shi, W., Cao, J., Zhang, Q., Li, Y., & Xu, L. 2016. Edge computing: Vision and challenges, *IEEE Internet of Things Journal*, 3(2016), 637–646.

45. Kaur, A., Gupta, P., Singh, M., & Nayyar, A. 2019. Data placement in era of cloud computing: A survey, taxonomy and open research issues. *Scalable Computing: Practice and Experience*, 20(2), 377–398.

46. Boeing 787s to Create Half a Terabyte of Data Per Flight, Says Virgin Atlantic. n.d. https://datafloq.com/read/self-driving-carscreate-2-petabytes-data-annually/172 (accessed December 7, 2021).

47. Shi, W., & Dustdar, S. 2016. The promise of edge computing. *Computer*, 49(5), 78–81.

48. Garcia Lopez, P., Montresor, A., Epema, D., Datta, A., Higashino, T., Iamnitchi, A., ... & Riviere, E. 2015. Edge-centric computing: Vision and challenges. *ACM SIGCOMM Computer Communication Review*, 45(5), 37–42.

49. Satyanarayanan, M., Simoens, P., Xiao, Y., Pillai, P., Chen, Z., Ha, K., ... & Amos, B. 2015. Edge analytics in the internet of things. *IEEE Pervasive Computing*, 14(2), 24–31.

50. Rodrigues, T. G., Suto, K., Nishiyama, H., & Kato, N. 2016. Hybrid method for minimizing service delay in edge cloud computing through VM migration and transmission power control. *IEEE Transactions on Computers*, 66(5), 810–819.

51. Zhang, Y., Ren, J., Liu, J., Xu, C., Guo, H., & Liu, Y. 2017. A survey on emerging computing paradigms for big data. *Chinese Journal of Electronics*, 26(1), 1–12.

52. Ren, J., Guo, H., Xu, C., & Zhang, Y. 2017. Serving at the edge: A scalable IoT architecture based on transparent computing. *IEEE Network*, 31(5), 96–105.

53. Liu, J., Guo, H., Fadlullah, Z. M., & Kato, N. 2016. Energy consumption minimization for FiWi enhanced LTE-A HetNets with UE connection constraint. *IEEE Communications Magazine*, 54(11), 56–62.

54. Liu, J., Guo, H., Nishiyama, H., Ujikawa, H., Suzuki, K., & Kato, N. 2015. New perspectives on future smart FiWi networks: Scalability, reliability, and energy efficiency. *IEEE Communications Surveys & Tutorials*, 18(2), 1045–1072.

55. Li, H., Ota, K., & Dong, M. 2018. Learning IoT in edge: Deep learning for the Internet of Things with edge computing. *IEEE Network*, 32(1), 96–101.

56. Singh I., Singh K. S., Kumar S., & Aggarwal K. 2021. Dropout-VGG based convolutional neural network for traffic sign categorization. In *Proceeding of 2nd Congress on Intelligent Systems (CIS 2021)*, Springer Book Series on Data Engineering and Communication Technologies.

57. Bonomi, F., Milito, R., Zhu, J., & Addepalli, S. 2012. Fog computing and its role in the internet of things. In *Proceedings of the First Edition of the MCC Workshop on Mobile Cloud Computing*, 13–16.

58. Singh, S. P., Nayyar, A., Kaur, H., & Singla, A. 2019. Dynamic task scheduling using balanced VM allocation policy for fog computing platforms. *Scalable Computing: Practice and Experience*, 20(2), 433–456.

59. Singh, S. P., Kumar, R., Sharma, A., & Nayyar, A. 2020. Leveraging energy-efficient load balancing algorithms in fog computing. *Concurrency and Computation: Practice and Experience*, 34, e5913.

60. Roman, R., Lopez, J., & Mambo, M. 2018. Mobile edge computing, fog et al.: A survey and analysis of security threats and challenges. *Future Generation Computer Systems*, 78, 680–698.

61. Gusev, M., & Dustdar, S. 2018. Going back to the roots—The evolution of edge computing, an iot perspective. *IEEE Internet Computing*, 22(2), 5–15.

62. Li, C., Xue, Y., Wang, J., Zhang, W., & Li, T. 2018. Edge-oriented computing paradigms: A survey on architecture design and system management. *ACM Computing Surveys (CSUR)*, 51(2), 1–34.

63. Chiang, M., Ha, S., Risso, F., Zhang, T., & Chih-Lin, I. 2017. Clarifying fog computing and networking: 10 questions and answers. *IEEE Communications Magazine*, 55(4), 18–20.

64. Neto, P. 2011. Demystifying cloud computing. In *Proceedings of the Doctoral Symposium on Informatics Engineering*, 24, 16–21. Porto, Portugal.

65. Singh S., Kaur K., & Aggrwal A. 2014. Emerging trends and limitations in technology and system of ubiquitous computing. *International Journal of Advanced Research in Computer Science (IJARCS)*, 5(7), 174–178.

66. Singh S., Aggrwal A., & Kaur K. 2015. Evaluation & trends of surveillance system network in ubiquitous computing environment. *Int. J. Advanced Networking and Application (IJANA)*, 6(5), 2486–2493.

67. Liu, F., Mao, J., & Tong, J. 2011. NIST Cloud Computing Reference Architecture Version 1. *National Institute of Standards and Technology*.

68. Nayyar, A. 2019. *Handbook of Cloud Computing: Basic to Advance Research on the Concepts and Design of Cloud Computing*. BPB Publications.

69. Singh, S. 2016. Empowering high-performance computing over cloud, cluster & grid computing. Network. *Journal of Multidisciplinary Engineering Technologies (JMDET)*, 10(2), 6–13.

70. Roman, R., Lopez, J., & Mambo, M. 2018. Mobile edge computing, fog et al.: A survey and analysis of security threats and challenges. *Future Generation Computer Systems*, 78, 680–698.

71. Cao, K., Liu, Y., Meng, G., & Sun, Q. 2020. An overview on edge computing research. *IEEE Access*, 8, 85714–85728.

72. Saharan, K. P., & Kumar, A. 2015. Fog in comparison to cloud: A survey. *International Journal of Computer Applications*, 122(3).

73. Margariti, S. V., Dimakopoulos, V. V., & Tsoumanis, G. 2020. Modeling and simulation tools for fog computing—A comprehensive survey from a cost perspective. *Future Internet*, 12(5), 89.

74. Khan, W. Z., Ahmed, E., Hakak, S., Yaqoob, I., & Ahmed, A. 2019. Edge computing: A survey. *Future Generation Computer Systems*, 97, 219–235.

75. Salman, O., Elhajj, I., Chehab, A., & Kayssi, A. 2018. IoT survey: An SDN and fog computing perspective. *Computer Networks*, 143, 221–246.

76. Tiwary, M., Puthal, D., Sahoo, K. S., Sahoo, B., & Yang, L. T. 2018. Response time optimization for cloudlets in mobile edge computing. *Journal of Parallel and Distributed Computing*, 119, 81–91.

77. Cao, B., Sun, Z., Zhang, J., & Gu, Y. 2021. Resource allocation in 5G IoV architecture based on SDN and fog-cloud computing. *IEEE Transactions on Intelligent Transportation Systems*, 22(6), 3832–3840.

78. Sahoo, K. S., Tiwary, M., Luhach, A. K., Nayyar, A., Choo, K. K. R., & Bilal, M. 2021. Demand-supply based economic model for resource provisioning in industrial IoT traffic. *IEEE Internet of Things Journal*.

79. Chekired, D. A., Togou, M. A., & Khoukhi, L. 2018. A hybrid SDN path computation for scaling data centers networks. In *2018 IEEE Global Communications Conference (GLOBECOM)*, 1–6. IEEE.

80. Sahoo, K. S., Mishra, P., Tiwary, M., Ramasubbareddy, S., Balusamy, B., & Gandomi, A. H. 2019. Improving end-users utility in software-defined wide area network systems. *IEEE Transactions on Network and Service Management*, 17(2), 696–707.

81. Ibrar, M., Wang, L., Muntean, G. M., Chen, J., Shah, N., & Akbar, A. 2020. IHSF: An intelligent solution for improved performance of reliable and time-sensitive flows in hybrid SDN-based FC IoT systems. *IEEE Internet of Things Journal*, 8(5), 3130–3142.

82. Zemrane, H., Baddi, Y., & Hasbi, A. 2018. SDN-based solutions to improve IoT: Survey. In *2018 IEEE 5th International Congress on Information Science and Technology (CiSt)*, 588–593. IEEE.

83. Khan, L. U., Yaqoob, I., Tran, N. H., Kazmi, S. A., Dang, T. N., & Hong, C. S. 2020. Edge-computing-enabled smart cities: A comprehensive survey. *IEEE Internet of Things Journal*, 7(10), 10200–10232.

84. Arasteh, H., Hosseinnezhad, V., Loia, V., Tommasetti, A., Troisi, O., Shafie-Khah, M., & Siano, P. 2016. Iot-based smart cities: A survey. In *2016 IEEE 16th International Conference on Environment and Electrical Engineering (EEEIC)*, 1–6. IEEE.
85. Krishnamurthi, R., Nayyar, A., & Solanki, A. 2019. Innovation Opportunities through Internet of Things (IoT) for Smart Cities. In *Green and Smart Technologies for Smart Cities*, 261–292. CRC Press.
86. Chakrabarty, S., & Engels, D. W. 2016. A secure IoT architecture for smart cities. In *2016 13th IEEE Annual Consumer Communications & Networking Conference (CCNC)*, 812–813. IEEE.
87. Haller, S., & Magerkurth, C. 2011. The real-time enterprise: Iot-enabled business processes. In *IETF IAB Workshop on Interconnecting Smart Objects with the Internet*, 1–3.
88. Zhang, Y., & Wen, J. 2017. The IoT electric business model: Using blockchain technology for the internet of things. *Peer-to-Peer Networking and Applications*, 10(4), 983–994.
89. Kodali, R. K., Swamy, G., & Lakshmi, B. 2015. An implementation of IoT for healthcare. In *2015 IEEE Recent Advances in Intelligent Computational Systems (RAICS)*, 411–416. IEEE.
90. Rohokale, V. M., Prasad, N. R., & Prasad, R. 2011. A cooperative Internet of Things (IoT) for rural healthcare monitoring and control. In *2011 2nd international conference on wireless communication, vehicular technology, information theory and aerospace & electronic systems technology (Wireless VITAE)*, 1–6. IEEE.
91. Chandy, A. 2019. A review on iot based medical imaging technology for healthcare applications. *Journal of Innovative Image Processing (JIIP)*, 1(1), 51–60.
92. Al-Emran, M., Malik, S. I., & Al-Kabi, M. N. 2020. A Survey of Internet of Things (IoT) in Education: Opportunities and Challenges. In *Toward Social Internet of Things (SIoT): Enabling Technologies, Architectures and Applications*, 197–209.
93. Marquez, J., Villanueva, J., Solarte, Z., & Garcia, A. 2016. IoT in Education: Integration of Objects with Virtual Academic Communities. In *New Advances in Information Systems and Technologies*, 201–212. Springer, Cham.
94. Ahmed, N., De, D., & Hussain, I. 2018. Internet of Things (IoT) for smart precision agriculture and farming in rural areas. *IEEE Internet of Things Journal*, 5(6), 4890–4899.
95. Shenoy, J., & Pingle, Y. 2016. IOT in agriculture. In *2016 3rd International Conference on Computing for Sustainable Global Development (INDIACom)*, 456–1458. IEEE.
96. Stočes, M., Vaněk, J., Masner, J., & Pavlík, J. 2016. Internet of things (iot) in agriculture-selected aspects. *Agris On-Line Papers in Economics and Informatics*, 8(665-2016-45107), 83–88.
97. Santoso, F. K., & Vun, N. C. 2015. Securing IoT for smart home system. In *2015 International Symposium on Consumer Electronics (ISCE)*, 1–2. IEEE.
98. Malche, T., & Maheshwary, P. 2017. Internet of Things (IoT) for building smart home system. In *2017 International Conference on I-SMAC (IoT in Social, Mobile, Analytics and Cloud) (I-SMAC)*, 65–70. IEEE.
99. Kumar, S., Singh, S. K., Aggarwal, N., & Aggarwal, K. 2021. Evaluation of automatic parallelization algorithms to minimize speculative parallelism overheads: An experiment. *Journal of Discrete Mathematical Sciences and Cryptography*, 24(5), 1517–1528.
100. Fernández, C. M., Rodríguez, M. D., & Muñoz, B. R. 2018. An edge computing architecture in the Internet of Things. In *2018 IEEE 21st International Symposium on Real-Time Distributed Computing (ISORC)*, 99–102. IEEE.
101. Mukherjee, M., Matam, R., Shu, L., Maglaras, L., Ferrag, M. A., Choudhury, N., & Kumar, V. 2017. Security and privacy in fog computing: Challenges. *IEEE Access*, 5, 19293–19304.
102. Schaberg, M. 2017. Partnering on your Journey. KLM Services, LLC. https://www.klmservices.com/mtcontent/uploads/2017/12/our-journey-presentation-osme-120417.pdf (accessed on December 20, 2021).
103. Pfitzmann, A., & Hansen, M. 2008. Anonymity, unlinkability, undetectability, unobservability, pseudonymity, and identity management-a consolidated proposal for terminology. *Version v0*, 31, 15.
104. Zhang, J., Chen, B., Zhao, Y., Cheng, X., & Hu, F. 2018. Data security and privacy-preserving in edge computing paradigm: Survey and open issues. *IEEE Access*, 6, 18209–18237.

105. Archana, B. S., Chandrashekar, A., Bangi, A. G., Sanjana, B. M., & Akram, S. 2017. Survey on usable and secure two-factor authentication. In *2017 2nd IEEE International Conference on Recent Trends in Electronics, Information & Communication Technology (RTEICT)*, 842–846. IEEE.

106. Authy. n.d. What is two-factor authentication (2fa)? https://authy.com/what-is-2fa/ (accessed on December 21, 2021).

107. Varghese, B., Wang, N., Barbhuiya, S., Kilpatrick, P., & Nikolopoulos, D. S. 2016. Challenges and opportunities in edge computing. In *2016 IEEE International Conference on Smart Cloud (SmartCloud)*, 20–26. IEEE.

108. Satyanarayanan, M., Bahl, P., Caceres, R., & Davies, N. 2009. The case for vm-based cloudlets in mobile computing. *IEEE pervasive Computing*, 8(4), 14–23.

109. Khanna, A., & Anand, R. 2016. IoT based smart parking system. In *2016 International Conference on Internet of Things and Applications (IOTA)*, 266–270. IEEE.

4

An Overview of Software-Defined Internet of Vehicles

Sagarika Mohanty, Bibhudatta Sahoo and Hemant Kumar Apat
National Institute of Technology, Rourkela, India

Kshira Sagar Sahoo
SRM University, Amaravati, India
Umeå° University, Umeå° 901 87, Sweden

CONTENTS

4.1 Introduction

Internet of Vehicles (IoV) is the confluence of the internet of things (IoT) and the mobile internet, in which automobiles act as intelligent objects, and has become a cornerstone of smart cities and intelligent transportation systems (ITS). Safety, traffic efficiency and entertainment are usually the focus of IoV applications. The IoV's brand new technological frontier has become a reality due to the rapid advance of vehicle technology, including numerous sensors, radio transceivers, computing units, the enhancement of roadside units, electronic processing units, communications and storage capacity [1].

Using the software-defined networking (SDN) methodology, the software-defined internet of vehicles (SD-IoV) has been developed as a method of enhancing IoV performance and management by achieving improved network programmability obtained by the separation of control and infrastructure planes [2]. It brings smartness to the automotive environment by integrating vehicular networks with learning from the internet of things, enhancing vehicle intelligence and networking. It aims to connect individuals, cars and devices in a worldwide network that connects disparate networks and provides services for smart transportation.

The automotive environment includes unique characteristics, such as high dynamic topology, large amounts of data, frequent disconnections and specific needs, including high quality of service, reliable communications and low latency [3–5]. Cloud computing is used in vehicular communications to provide centralized compute and storage services. Data may be obtained from any location without requiring a great amount of storage or computational power in the vehicles. The time delay between transmitting data from vehicles to the cloud server and obtaining the information after it has been saved and analyzed is a significant challenge. Edge computing is a better alternative since it allows data to be processed and analyzed closer to the end devices. The edge acts as a gateway between the cloud and the automobiles. Edge computing provides a higher quality of service (QoS) since computing and storage services are delivered near to the customer (on the edge) [6–9]. SDN accompanies edge computing and supports the intelligent transportation system to enable instantaneous vehicle mobility and promote safe driving despite the complexity of a heterogeneous and large number of automobiles, massive data flow, and rapid topology changes [10]. To manage these challenges, SDN has been suggested as a viable alternative.

SDN is a revolutionary network architecture that transforms existing networks into programmable networks. Among its many advantages including network programmability and infrastructure abstraction, which provides an efficient network through centralized control [11]. SDN's flexibility, centralization and programmability can be used to enhance network optimization, resource usage and quality of service [12, 13]. A new field known as software-defined internet of vehicles (SD-IoV) has emerged, combining SDN and IoV.

This chapter provides background information on SDN, VANETs and IoV, explains the general and the layered architecture of SD-IoV, and discusses some current issues. The remainder of the chapter is organized as follows: Section 4.2 provides background information on the SD-IoV architecture, including software-defined networking and internet of vehicles. Section 4.3 provides an overview of software-defined internet of vehicles. The SD-IoV architecture is described in Section 4.4. Section 4.5 covers various open issues and Section 4.6 concludes the chapter with future scope.

4.2 Background

4.2.1 Software-Defined Networks (SDN)

SDN is a modern architecture that separates network control and forwarding devices. This eliminates the need for manual intervention and simplifies the management of the network.

SDN has gained popularity in recent times, in the context of several next-generation networking technologies, including 5G, the internet of things, named data networks (NDNs), vehicular ad-hoc networks (VANETs) and sensor networks [14].

The control plane in SDN serves as a centralized unit to manage the various components of a network such as network traffic and its forwarding. Due to the increasing importance of networking in various next-generation networks, the term SDN has also gained increasing importance. Figure 4.1 highlights the architecture of SDN.

4.2.1.1 Infrastructure Layer

The data plane or infrastructure consists of routers, switches (physical or virtual) and access points which are the most important components in this tier. These devices use an open interface to execute packet routing, switching and forwarding. Separate secure channels are employed to control connections to network devices and for user data flows.

4.2.1.2 Control Layer

In SDN, the control layer or the control plane sits between the application and data planes. It comprises a series of SDN controllers that maintain a global and dynamic network view, accept application layer requests, and manage network devices using standard protocols.

4.2.1.3 Application Layer

The application layer, also known as the management plane, includes network services and end-user business applications. Management systems, monitoring the network, flow control and security are some examples of business applications.

To obtain network state information from the infrastructure layer, an application programming interface (API) is used by the SDN applications.

- **Southbound Interface**
  This interface enables connection between the SDN controller and the infrastructure layer's forwarding elements. The communication follows a common protocol. The most widely used is OpenFlow, an open networking foundation (ONF).

- **Northbound Interface**
  The northbound interface connects the services and applications of the management plane to the controllers that reside in the control plane. This interface is made up of different APIs that provide programmability and are used to regulate traffic flows in the network dynamically.

- **Eastbound/Westbound Interface**
  This interface protocol is used to coordinate connection between controllers. The implementation of this interface is reliant on the underlying network technology because there are no clear standards for it. This interface is an example of a link between an SDN domain and a legacy domain that responds to message requests using routing protocol.

4.2.2 OpenFlow

Messages from controllers to data plane devices are sent through the southbound interface, which uses the OpenFlow communications protocol. To facilitate variable network traffic control, OpenFlow provides access to router flow tables [15]. The protocol manages tasks like topology setting and filtering of packets.

There are several benefits of SDN, such as centralized network provisioning, higher rate of innovation, increased network reliability and security, hardware savings and reduced capital expenditure, lower operating costs, cloud abstraction, more granular network control, and better user experience [16–20].

4.2.3 Network Function Virtualization (NFV)

Network function virtualization delegates network services such as firewalls and domain name service (DNS) to an application that runs on a standardized infrastructural unit including storage units, switches and servers. Both wired and wireless networks can benefit from NFV. NFV lowers capital (CAPEX) and operational costs (OPEX) by reducing the need for specific hardware devices and their power generation units and coolants. The use of a software program running on servers to virtualize different network functions enables rapid scale-up and scale-down of different services as well as flexible delivery of these services. By combining NFV with SDN, expensive and dedicated hardware equipment can be replaced with software and generic hardware [15].

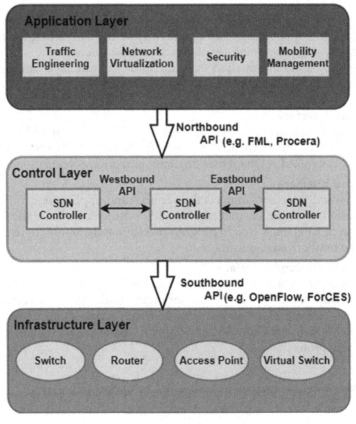

FIGURE 4.1
Architecture of SDN.

4.3 Internet of Vehicles (IoV)

The primary goal of intelligent transportation systems is to enhance traffic efficiency, road safety and passenger information. VANETs (vehicular ad-hoc networks) have evolved to provide a variety of data services and applications for ITS. The purpose is to provide traffic and safety information as well as emergency notifications to drivers in real time. The general architecture of a VANET is depicted in Figure 4.2 [21].

In VANETs, vehicles and roadside units (RSUs) can communicate in three ways: V2V, V2R and inter-roadside communications (R2R). Though VANETs deliver traffic and safety information at a lower cost and in less time, there are a number of difficulties in modern vehicular communications that must be solved [22, 23]. VANET's ad-hoc network architecture, lack of commercialization, unstable internet connectivity, limited processing capability, incompatibility with personal devices, and lack of cloud computing services are among the issues. Besides this, there is a need for programmability, flexibility, and resource management in the VANET ecosystem [24–26].

The internet of things (IoT) is a new network structure that enables device components to communicate with one another, allowing them to become more informative, intelligent, and smarter, resulting in a vision of "anytime, anywhere, anything" communications [27].

Smart cities and homes, healthcare, industry, transportation, energy and agriculture are all examples of IoT, which is an amalgamation of heterogeneous networks that has included

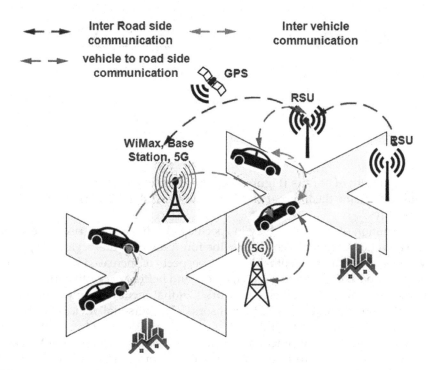

FIGURE 4.2
VANET architecture [21].

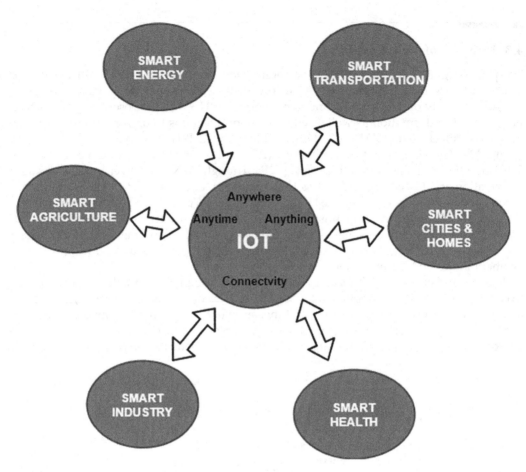

FIGURE 4.3
Impact of IoT on various research areas.

smartness in a variety of sectors (Figure 4.3). The internet of vehicles is a new domain in ITS that aims to combine the internet of things with vehicular ad-hoc networks to improve their functionality.

The amalgamation of IoT technology in specific fields have given rise to new concepts. Industry 4.0 and the internet of vehicles are the forerunners in this regard [28]. The future of the IoV appears bright and profitable with prospects of increased road safety, reduced environmental impact, efficient cost management and better space utilization.

The internet of vehicles is a diverse network [29] that includes communication among vehicles (V2V), with roadside units (V2R), sensors (V2S), personal devices (V2P) and infrastructure (V2I) as illustrated in Figure 4.4. It is an interconnected network of automobiles, cellphones and wearables, equipment and networks. Each vehicle in the IoV is an intelligent object with sensors, control units and numerous computing facilities that can communicate with any entity via a V2X communication architecture [30].

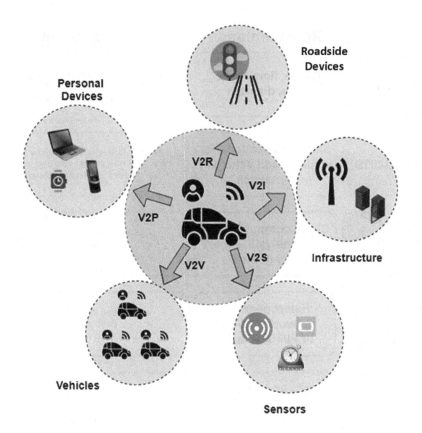

FIGURE 4.4
Types of communications in IoV.

The goal of IoV is to automate numerous security and efficiency feature in vehicles, as well as to commercialize vehicular networks (Figure 4.5).

- *Perception layer:* This layer uses equipment such as sensors, actuators, wearables, cellphones, and personal devices that are already installed in the vehicles as well as roadside units (RSUs) and other intelligent infrastructure components. One of the most significant component of this layer is the global ID (GID) terminal, which provides the vehicle with a recognizable ID and performs radio-frequency identification (RFID) functions. This layer's main objective is to describe the interaction between the vehicle and its environment as well as to collect data on events, the environment, driving behaviors and patterns in order to acquire useful data for use in vehicle decisions.

- *Coordination layer:* Interoperability, routing and message transportation, security, etc. are handled by this layer. The primary focus is to reassemble the various structures of data received from various networks including WAVE, 4G/LTE, Wi-Fi and satellite networks into a single network structure. It guarantees that all networks are connected.

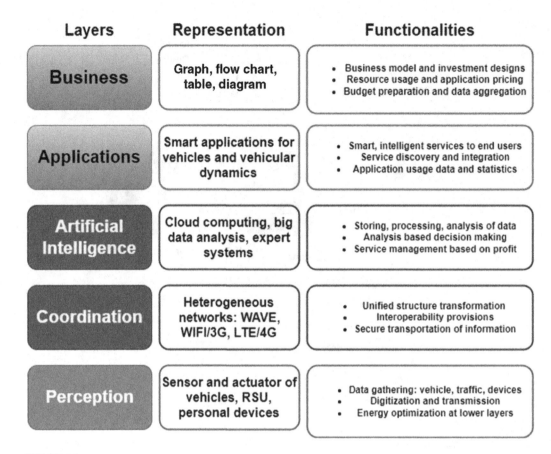

FIGURE 4.5
Layered architecture of IoV.

- *Artificial intelligence (AI) layer:* This layer is in charge of managing the storage, processing and analysis of the data as well as taking decisions after doing significant analysis. It largely focuses on data mining, cloud computing, big data analytics and expert-based decision making. This also manages service management which is crucial aspect for IoV.

- *Application layer:* Traffic control, safety, multimedia-based and web-enabled smart applications make up the fourth layer, which is in charge of commercializing IoV technology. The layer is in charge of providing end users with intelligent and critical analysis of available data, based on this layer's smart services.

- *Business layer:* The key focus of this layer is to generate business model and investment strategies based on statistical application usage and data analysis. For doing these statistical analyses different tools are used, such as comparison tables, graphs, use-case diagrams and flowcharts. Investment and resource management decisions, charging application usage, budget preparation and data aggregation are some other tasks.

There are several benefits of IoV such as enhanced safety, faster travel and convenience, reduced congestion on the roads, optimized routes, better parking and remote car management.

4.3.1 Software-Defined Internet of Vehicles (SD-IoV)

Software-defined networking and network function virtualization are complementary approaches for designing and managing networks. SDN technology provides a platform for exploring and implementing new concepts, as well as examining their programmability and centralized management. The data plane and control plane are separated, allowing for a centralized network view [32].

The internet of vehicles is a new technology that is fast evolving. Government agencies, industries, and researchers have made significant efforts to develop and deploy an efficient vehicular communication system that will noticeably contribute to the deployment of intelligent transportation systems. IoV has unique qualities such as high processing capability, high-speed internet access, mobility and varying network density. The internet of vehicles differs from vehicular ad-hoc networks in that it has centralized management, making it ideal for networks that allow vehicles on the road to connect to each other.

In the United States, an additional 20.8 million vehicles are expected to be sold by 2030. Traffic control and road safety applications on such a large scale is a difficult endeavor for VANETs. VANETs require a programmable architecture to bridge this gap and provide current transportation services.

The internet of vehicles network records data about vehicles such as direction, speed, and route, etc. Cars can collect information about their surroundings and states through the global positioning system, sensors, camera, RFID and other technologies. Through the internet they can send their varied data to the central processing unit. These massive data can be examined and processed using a variety of approaches to find the best route for different vehicles as well as provide real-time road conditions and traffic-light cycles [33, 34].

The internet of vehicles network system allows vehicles, roadside units and the internet to communicate and exchange information wirelessly. It consists of intra- and inter-vehicular communications and vehicular mobile internet, all of which adhere to specific protocols and standards for communications [35, 36]. By processing large volumes of data, the system generates an integrated network of smart traffic control, real-time information and vehicle control.

The internet of things, cloud technology and big data are all used by the IoV system to provide users with information about traffic efficiency, safety and infotainment. The data collected from vehicles is uploaded to the central processing unit, which analyzes and processes it and provides it to the user [37, 38].

ITS requires vehicle-to-vehicle and vehicle-to-roadside (V2V and V2R) connectivity with varying levels of service quality (QoS) to improve user satisfaction. It is essential to keep track of the growing number of connected vehicles which are also being able to handle a vast number of concurrent queries. For doing this in both wired and wireless networks, software-defined networking is a solution. SDN can improve flexibility and efficiency and create a platform for advanced network management because of the separation of the control and data planes [39–41]. SD-IoV (software-defined IoV) integrates SDN and IoV by gathering network status information and making relevant decisions [42]. The controllers in SD-IoV handle vehicle communications in a centralized manner. Allocation of resources may be optimized dynamically and quality of service can also be guaranteed using real-time global knowledge [43, 44].

4.4 Software-Defined IoV: Architecture

Figure 4.6 depicts a generalized architecture of SD-IoV. The major components and their functionalities are represented as follows:

- *SDN controllers:* These are software programs running in servers and providing a global view in the SD-IoV system. They are in charge of network operation and management which includes creation of rules, association of clients, mobility management and virtualization of the network as well as some advanced features like data pre-processing, network analysis, and learning.
- *SDN switches:* The operator manages the SDN switch network which is linked to the internet via high speed. Similar to Google's software-defined WAN (wide area

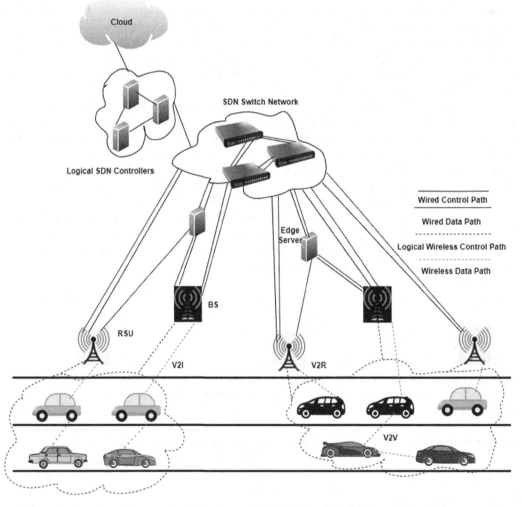

FIGURE 4.6
Architecture of SD-IoV.

network), it can expand to as large as a global network, depending on the operator's demand.

- *SDN-enabled wireless access infrastructures:* Cellular base stations, Wi-Fi access points, roadside units and DSA coordinators can be used to offer wireless interfaces to the linked vehicles. The infrastructures are SDN enabled and execute tasks including packet forwarding and transmission as well as infrastructure-specific functions.

   RSUs and base stations (BSs) are the edge devices. The edge server serves as a vehicle storage and processing facility at the network's edge, storing regional road system data and providing emergency services [45–48].

- *SDN-enabled vehicle on-board units (OBUs):* Vehicles/automobiles are the end users in the SD-IoV system. Packet forwarding, selection of channel, selection of interface, power control and selection of transmission mode are all SDN-enabled OBU tasks.

- *Data path:* This comprises wired and wireless connections. Data is transmitted through wired links between SDN-enabled switch networks and wireless access infrastructures. Wireless communications provide data exchange services for SDN-enabled automobiles.

- *Control path:* The major functions of the control paths are getting real-time state feedbacks from the components and controlling them. Wired control paths are used between controllers, switches and wireless access infrastructures (all SDN enabled). Wireless control paths can be implemented as per the requirements on control channels, protocols and hardware supports.

Figure 4.7 present the layered architecture of SD-IoV. All network services are located on the application plane, which is built on top of the controllers to materialize services and translate them into rules using various APIs. SDN-enabled switches and wireless infrastructures make up the upper data plane. The control plane defines the rules using the OpenFlow protocol. Specifically, due to the involvement of wireless access infrastructures, it is desirable to extend OpenFlow to achieve more advanced functions than simple packet forwarding [49]. SDN-enabled vehicles are the end users and sometimes act as relays (i.e., in V2V communications) form the lower data plane. The application plane and control plane require current-state information from the data planes in order to generate rules.

4.5 Software-Defined IoV: Open Issues

The following are some of the open issues connected to Software-Defined IoV:

- **Resource Abstraction**
  Abstraction of physical resources is one approach to diminish the problem of heterogeneous resources of SD-IoV. If uniform language is used for the heterogeneous resource then resource scheduling will be easier and more efficient use of programmability as offered by SDN [44].

- **Network Intelligence**
  SD-IoV is inherently appropriate for realizing an intelligent network, which means the controller has the ability to dynamically optimize the network resource management through learning-based algorithms [50].

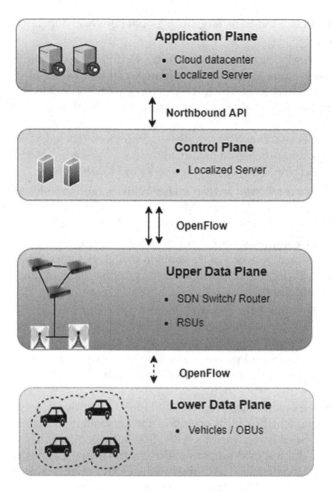

FIGURE 4.7
Layered architecture of SD-IoV.

- **Resource Management**
  The collection, storage and analysis of all data from connected vehicles, as well as resource management, is a huge undertaking. SD-IoV manages resources efficiently and shares resources to fulfill user demands in the situation of dynamic topology, inconsistent connectivity, or restricted storage capacity of local servers [41].

- **Security and Privacy**
  As a centrally controlled and location-based system, SD-IoV will encounter various security and privacy challenges. Controllers require detailed descriptions of vehicle status, such as current positions, speed and destinations. If the controller or its databases are compromised, the privacy of the vehicular end user will be jeopardized. Also, due to centralization, SD-IoV is vulnerable to DoS (denial-of-service) attacks. The control functionality may be paralyzed by a huge number of fake requests received within a short time window. Moreover, if the controller itself is seized by unauthorized people, they can utilize it for further malicious purposes such as initiating distributed DoS attacks to other networks or valuable data theft [44].

4.6 Conclusion and Future Scope

The internet of vehicles faces numerous hurdles because of the limited processing resources and capability of on-board technology. Edge computing and software-defined networking have the potential to improve IoV network dynamics, allowing applications to perform better. This chapter has first identified the need for SD-IoV, then provided background information on SDN, VANETs and IoV, which are the key components of SD-IoV architecture. Generalized SD-IoV architecture was presented, followed by layered architecture. Finally, some open issues associated with SD-IoV are discussed. In the future, the focus will be to address issues related to edge computing-enabled software-defined internet of vehicles.

Acknowledgment

The Department of Science and Technology is funding this research through the WOS-A scheme with file no.SR/WOS-A/ET-133/2017.

References

1. Ang, L. M., Seng, K. P., Ijemaru, G. K., & Zungeru, A. M. (2018). Deployment of IoV for smart cities: Applications, architecture, and challenges. *IEEE Access*, 7, 6473–6492.
2. Yuan, T., da Rocha Neto, W., Rothenberg, C. E., Obraczka, K., Barakat, C., & Turletti, T. (2020). Dynamic controller assignment in software defined Internet of Vehicles through multi-agent deep reinforcement learning. *IEEE Transactions on Network and Service Management*, 18(1), 585–596.
3. Hou, X., Ren, Z., Wang, J., Cheng, W., Ren, Y., Chen, K. C., & Zhang, H. (2020). Reliable computation offloading for edge-computing-enabled software-defined IoV. *IEEE Internet of Things Journal*, 7(8), 7097–7111.
4. Zhai, Y., Sun, W., Wu, J., Zhu, L., Shen, J., Du, X., & Guizani, M. (2020). An energy aware offloading scheme for interdependent applications in software-defined IoV with fog computing architecture. *IEEE Transactions on Intelligent Transportation Systems*, 22(6), 3813–3823.
5. Mohanty, S., Shekhawat, A. S., Sahoo, B., Apat, H. K., & Khare, P. (2021, December). Minimizing Latency for Controller Placement Problem in SDN. In *2021 19th OITS International Conference on Information Technology (OCIT)* (pp. 393–398). IEEE.
6. Lin, C., Han, G., Qi, X., Guizani, M., & Shu, L. (2020). A distributed mobile fog computing scheme for mobile delay-sensitive applications in SDN-enabled vehicular networks. *IEEE Transactions on Vehicular Technology*, 69(5), 5481–5493.
7. Zhang, Y., Zhang, H., Long, K., Zheng, Q., & Xie, X. (2018). Software-defined and fog-computing-based next generation vehicular networks. *IEEE Communications Magazine*, 56(9), 34–41.
8. Apat, H. K., Sahoo, B., & Mohanty, S. (2021, December). A Quality of Service (QoS) Aware Fog Computing Model for Intelligent (IoT) Applications. In *2021 19th OITS International Conference on Information Technology (OCIT)* (pp. 267–272). IEEE.
9. Neha, B., Panda, S. K., Sahu, P. K., Sahoo, K. S., & Gandomi, A. H. (2022). A systematic review on osmotic computing. *ACM Transactions on Internet of Things*, 3(2), 1–30.

10. Pokhrel, S. R. (2021). Software defined internet of vehicles for automation and orchestration. *IEEE Transactions on Intelligent Transportation Systems*.

11. Sahoo, K. S., Mohanty, S., Tiwary, M., Mishra, B. K., & Sahoo, B. (2016, August). A Comprehensive Tutorial on Software Defined Network: The Driving Force for the Future Internet Technology. In *Proceedings of the International Conference on Advances in Information Communication Technology & Computing* (pp. 1–6).

12. Wibowo, F. X., Gregory, M. A., Ahmed, K., & Gomez, K. M. (2017). Multi-domain software defined networking: Research status and challenges. *Journal of Network and Computer Applications*, 87, 32–45.

13. Bhatia, J., Dave, R., Bhayani, H., Tanwar, S., & Nayyar, A. (2020). SDN-based real-time urban traffic analysis in VANET environment. *Computer Communications*, 149, 162–175.

14. Das, T., Sridharan, V., & Gurusamy, M. (2019). A survey on controller placement in SDN. *IEEE Communications Surveys & Tutorials*, 22(1), 472–503.

15. Jammal, M., Singh, T., Shami, A., Asal, R., & Li, Y. (2014). Software defined networking: State of the art and research challenges. *Computer Networks*, 72, 74–98.

16. Mohanty, S., Priyadarshini, P., Sahoo, S., Sahoo, B., & Sethi, S. (2019a, October). Metaheuristic Techniques for Controller Placement in Software-Defined Networks. In *TENCON 2019–2019 IEEE Region 10 Conference (TENCON)* (pp. 897–902). IEEE.

17. Mohanty, S., Priyadarshini, P., Sahoo, B., & Sethi, S. (2019b, March). A Reliable Capacitated Controller Placement in Software Defined Networks. In *2019 3rd International Conference on Computing Methodologies and Communication (ICCMC)* (pp. 822–827). IEEE.

18. Mohanty, S., Kanodia, K., Sahoo, B., & Kurroliya, K. (2020, February). A Simulated Annealing Strategy for Reliable Controller Placement in Software Defined Networks. In *2020 7th International Conference on Signal Processing and Integrated Networks (SPIN)* (pp. 844–849). IEEE.

19. Sahoo, K. S., & Puthal, D. (2020). SDN-assisted DDoS defense framework for the internet of multimedia things. *ACM Transactions on Multimedia Computing, Communications, and Applications (TOMM)*, 16(3s), 1–18.

20. Rout, S., Sahoo, K. S., Patra, S. S., Sahoo, B., & Puthal, D. (2021). Energy efficiency in software defined networking: A survey. *SN Computer Science*, 2(4), 1–15.

21. Belamri, F., Boulfekhar, S., & Aissani, D. (2021). A survey on QoS routing protocols in Vehicular Ad Hoc Network (VANET). *Telecommunication Systems*, 1–37.

22. Mohanty, S., & Jena, D. (2012). Secure data aggregation in vehicular-adhoc networks: A survey. *Procedia Technology*, 6, 922–929.

23. Mohanty, S., Jena, D., & Panigrahy, S. K. (2012, August). A Secure RSU-Aided Aggregation and Batch-Verification Scheme for Vehicular Networks. In *International Conference on Soft Computing and its Applications (ICSCA2012)*, pp. 174–178.

24. Cardona, N., Coronado, E., Latré, S., Riggio, R., & Marquez-Barja, J. M. (2020). Software-defined vehicular networking: opportunities and challenges. *IEEE Access*.

25. Yaqoob, I., Ahmad, I., Ahmed, E., Gani, A., Imran, M., & Guizani, N. (2017). Overcoming the key challenges to establishing vehicular communication: Is SDN the answer?. *IEEE Communications Magazine*, 55(7), 128–134.

26. Mohanty, S., Sahoo, K. S., Sethi, S., & Sahoo, B. A data scheduling approach for software defined Vehicular Ad-Hoc Networks. *International Journal of Computer Applications*, 975, 8887.

27. Krishnamurthi, R., Kumar, A., Gopinathan, D., Nayyar, A., & Qureshi, B. (2020). An overview of IoT sensor data processing, fusion, and analysis techniques. *Sensors*, 20(21), 6076.

28. Sahoo, K. S., Tiwary, M., Luhach, A. K., Nayyar, A., Choo, K. K. R., & Bilal, M. (2021). Demand-supply based economic model for resource provisioning in industrial IoT traffic. *IEEE Internet of Things Journal*.

29. Sharma, S., & Kaushik, B. (2019). A survey on internet of vehicles: Applications, security issues & solutions. *Vehicular Communications*, 20, 100182.

30. Shen, X., Fantacci, R., & Chen, S. (2020). Internet of vehicles [scanning the issue]. *Proceedings of the IEEE*, 108(2), 242–245.

31. Kaiwartya, O., Abdullah, A. H., Cao, Y., Altameem, A., Prasad, M., Lin, C. T., & Liu, X. (2016). Internet of vehicles: Motivation, layered architecture, network model, challenges, and future aspects. *IEEE Access*, 4, 5356–5373.

32. Abbas, M. T., Muhammad, A., & Song, W. C. (2020). SD-IoV: SDN enabled routing for internet of vehicles in road-aware approach. *Journal of Ambient Intelligence and Humanized Computing*, 11(3), 1265–1280.

33. Sennan, S., Ramasubbareddy, S., Balasubramaniyam, S., Nayyar, A., Kerrache, C. A., & Bilal, M. (2021a). MADCR: Mobility aware dynamic clustering-based routing protocol in internet of vehicles. *China Communications*, 18(7), 69–85.

34. Sennan, S., Ramasubbareddy, S., Balasubramaniyam, S., Nayyar, A., Kerrache, C. A., & Bilal, M. (2021b). MADCR: Mobility aware dynamic clustering-based routing protocol in internet of vehicles. *China Communications*, 18(7), 69–85.

35. Lv, Z., Lloret, J., & Song, H. (2021). Guest editorial software defined internet of vehicles. *IEEE Transactions on Intelligent Transportation Systems*, 22(6), 3504–3510.

36. Sahbi, R., Ghanemi, S., & Djouani, R. (2018, October). A Network Model for Internet of Vehicles Based on SDN and Cloud Computing. In *2018 6th International Conference on Wireless Networks and Mobile Communications (WINCOM)* (pp. 1–4). IEEE.

37. Ramasubbareddy, S., Ramasamy, S., Sahoo, K. S., Kumar, R. L., Pham, Q. V., & Dao, N. N. (2020). Cavms: Application-aware cloudlet adaption and vm selection framework for multi-cloudlet environment. *IEEE Systems Journal*.

38. Pande, S. K., Panda, S. K., Das, S., Sahoo, K. S., Luhach, A. K., Jhanjhi, N. Z., … & Sivanesan, S. (2021). A resource management algorithm for virtual machine migration in vehicular cloud computing. *Computers, Materials & Continua*, 67(2), 2647–2663.

39. Krishnan, P., Jain, K., Buyya, R., Vijayakumar, P., Nayyar, A., Bilal, M., & Song, H. (2021). MUD-based behavioral profiling security framework for software-defined IoT networks. *IEEE Internet of Things Journal*.

40. Nayak, R. P., Sethi, S., Bhoi, S. K., Sahoo, K. S., Jhanjhi, N., Tabbakh, T. A., & Almusaylim, Z. A. (2021). TBDDoSA-MD: Trust-based DDoS misbehave detection approach in software-defined vehicular network (SDVN). *Cmc-Computers Materials & Continua*, 69(3), 3513–3529.

41. Kumar, A., Krishnamurthi, R., Nayyar, A., Luhach, A. K., Khan, M. S., & Singh, A. (2021). A novel software-defined drone network (SDDN)-based collision avoidance strategies for on-road traffic monitoring and management. *Vehicular Communications*, 28, 100313.

42. Raja, G., Dhanasekaran, P., Anbalagan, S., Ganapathisubramaniyan, A., & Bashir, A. K. (2020, July). SDN-Enabled Traffic Alert System for IoV in Smart Cities. In *IEEE INFOCOM 2020-IEEE Conference on Computer Communications Workshops (INFOCOM WKSHPS)* (pp. 1093–1098). IEEE.

43. Chen, J., Zhou, H., Zhang, N., Xu, W., Yu, Q., Gui, L., & Shen, X. (2017). Service-oriented dynamic connection management for software-defined internet of vehicles. *IEEE Transactions on Intelligent Transportation Systems*, 18(10), 2826–2837.

44. Indira, K., Ajitha, P., Reshma, V., & Tamizhselvi, A. (2019, February). An Efficient Secured Routing Protocol for Software Defined Internet of Vehicles. In *2019 International Conference on Computational Intelligence in Data Science (ICCIDS)* (pp. 1–4). IEEE.

45. Kaur, A., Singh, P., & Nayyar, A. (2020). Fog Computing: Building a Road to IoT with Fog Analytics. In *Fog Data Analytics for IoT Applications* (pp. 59–78). Springer, Singapore.

46 Apat, H. K., Sahoo Compt, B., Bhaisare, K., & Maiti, P. (2019, December). An Optimal Task Scheduling Towards Minimized Cost and Response Time in Fog Computing Infrastructure. In *2019 International Conference on Information Technology (ICIT)* (pp. 160–165). IEEE.

47. Apat, H. K., Maiti, P., & Patel, P. (2020b, December). Review on QoS Aware Resource Management in Fog Computing Environment. In *2020 IEEE International Symposium on Sustainable Energy, Signal Processing and Cyber Security (iSSSC)* (pp. 1–6). IEEE.

48. Apat, H. K., Bhaisare, K., Sahoo, B., & Maiti, P. (2020a, March). Energy Efficient Resource Management in Fog Computing Supported Medical Cyber-Physical System. In *2020 International Conference on Computer Science, Engineering and Applications (ICCSEA)* (pp. 1–6). IEEE.

49. Jiacheng, C., Haibo, Z. H. O. U., Ning, Z., Peng, Y., Lin, G., & Sherman, S. X. (2016). Software defined internet of vehicles: Architecture, challenges and solutions. *Journal of Communications and Information Networks*, 1(1), 14–26.

50. Jain, A., & Nayyar, A. (2020). Machine Learning and Its Applicability in Networking. In *New Age Analytics* (pp. 57–79). Apple Academic Press.

5

IoT Architecture and Research Issues in the Smart City Environment

Shreeya Swagatika Sahoo
Siksha 'O' Anusandhan Deemed to be University, Bhubaneswar, India

Asit Sahoo
National Institute of Technology Rourkela, Rourkela, India

Bibhudatta Sahoo
National Institute of Technology, Rourkela, India

CONTENTS

DOI: 10.1201/9781003213871-5

5.1 Introduction

With the increase in urbanization, cities and megacities are emerging rapidly [1]. A megacity is an urban area with a population over 10 million. Rapid population increase has expanded city borders which spread and merge into neighboring urban areas to form megacities. By 2030, the world is expected to contain 43 megacities, the majority of which will be in developing countries [2]. Numerous benefits such as higher standards of living, conveniences such as water supplies, sewage treatment, residential and office buildings, transportation, health services and so on draw people from rural to urban areas. According to a survey report by URBANET, world population is expected to grow by 2–3% each year [3]. However, the growing population in the cities has resulted in exhaustion of natural resources, creating ecological and environmental problems and public disorder problems. Various issues such as air and water pollution, building efficient infrastructure, etc. need to be addressed in overcrowded urban areas [4]. Growing urbanization needs sustainable development to improve safety and environmental conditions for all urban populations. Such development ensures sustainable use of urban resources such as investment in green infrastructure, sustainable industries, recycling and environmental campaigns, pollution management, renewable energy, green public transportation, and water recycling and reclamation [5]. The main objective of a smart city is to improve the quality of life of the inhabitants. For example, traffic issues can be solved with interconnected vehicles; smart monitoring can assist with healthcare emergencies; and the energy wastage in urban areas can be minimized using a smart grid. Ubiquitous connection to the internet and cloud will enable smart cities to provide security and safety with respect to disasters and malicious activity. Distributed computing and cloud services can guide the user through the city using augmented reality. The smart city can also provide data security through the use of cloud storage. Citizens can take advantage of many such features to experience improved quality of life.

In 2022, the world is moving toward a socioeconomic and environmental crisis. As the urban population is exponentially increasing, so too is poverty, waste, and power requirements. A city that can manage to navigate this route efficiently using digital data and technology to improve the quality of life of urban residents is called a smart city. In many rapidly developing cities, the authorities have not managed to design effective resource use in urban areas. The introduction of information communication technologies (ICT) such as big data, the internet of things (IoT), cloud computing, mobile computing, etc., has played an important role in the evolution of cities [6]. The term "smart city" was coined in the 1990s by researchers targeting technology and innovation in the urbanization process [7]. Though a formal definition of a smart city exists [8], we propose a definition of the smart city as the integration of smart technologies to predict, plan, and control the utilization of urban resources while both satisfying the needs of residents and ensuring economic growth. South Korea, China, India, and Russia have developed the smart city projects Songdo IBD [9], Meixi Lake [10], Lavasa [11], and Skolkovo [12], respectively to achieve efficient management of their urban resources. Smart city applications that have been developed include smart street lighting [13], smart traffic management systems [14], virtual power plants [15], smart health [16], smart emergency systems for crime prevention, dealing with accidents and natural disasters [17], and so on.

A unified framework needs to be developed to achieve seamless resource data management. For instance, the smart integration of medicines and stakeholder management with seamless monitoring of the production date and production source of medications can limit the distribution of fake and out-of-date medicines in the cities. Similarly, smart integration of healthcare with the traffic system can ensure help reaches accident victims more rapidly.

The smart city consists of various categories of end users: residents, government agencies, and so on [6]. Each user category has its own requirements and applications in the smart city scenario. For instance, a resident in a smart home may require high-quality connectivity and less security for browsing social networking sites and controlling appliances than a smart healthcare system that involves secure communication to the servers for storage and retrieval of personal information. As applications in smart cities are not mutually exclusive, both security and interoperability challenges need to be addressed to ensure secure and seamless communication in the smart city environment. In recent years, to establish an intelligent infrastructure for cities, many technologies like sensors, efficient communication protocols, and storage mechanisms have been developed. Among these concepts, the internet of things (IoT) is considered the backbone of the framework.

An IoT is formed by the exchange of information between devices through an interconnected medium. The key difference between the internet and IoT is the choice and importance of information shared by various device users. Here, sharing of key information helps to achieve a common objective. Urban areas of the modern world contain billions of devices that aid user convenience. The capability and number of these devices are increasing exponentially every year. These devices, together with information sharing and processing, can contribute to the development of smart cities. The speed and security of IoT depends on the quality of end-to-end delivery between devices, which is achieved due to the development of cloud infrastructure. Resource-heavy applications have moved to the cloud, leaving the devices with little to no load. Moreover, edge computing and next-generation networking are enabling faster information sharing and processing. The ongoing development of various low-power communication protocols and security measures has increased the lifetime of IoT devices. Additionally, the improvement in operating procedures and standardization has opened the way for scalability with heterogeneous devices. And it is not impractical to assume that in the future, the quality of service will become even better.

The key feature of a smart city is to provide a sustainable livelihood for urban people through smart information sharing and a developed infrastructure. In the current scenario, the technology is more focused on developing a faster and efficient information-sharing system. However, little to no focus has been applied to the sustainable utilization of resources and human value. This study offers a detailed analysis of the current scenario and the future roadmap for IoT in a smart city.

Organization of the Chapter

Section 5.2 provides an overview of existing smart city architecture systems. In Section 5.3, current technology available for the smart city environment is highlighted. Sections 5.4 and 5.5 demonstrate the smart city scenario and smart city aspect, respectively. Section 5.6 elaborates on issues related to smart cities. Section 5.7 shows the usefulness and challenges of the smart city in India. Finally, Section 5.8 concludes the chapter.

5.2 Related Work

Several researchers have developed IoT architecture for the smart city for a specific set of applications such as waste management [18], air quality monitoring [19], noise pollution [20], and so on. However, a standardized architecture that can address all smart city applications is required. Several use cases that belong to various application domains, such as healthcare management, smart mobility, residents' security, efficient farming, [21, 22], etc., need to be taken into consideration when designing a multi-level system. The architecture should be capable of sensing data from the city, transferring the aggregated data to be used by data analysis tools, providing information on the visual application tool and monitoring the functionalities of the solutions. The architecture needs to be flexible to support the aggregation of new products into the system. The architecture configuration needs to support remote management of devices. Data virtualization needs to be enabled to allow IoT devices to be set up with no limit on the number of devices in the system. Further, the system should satisfy security requirements and should use end-to-end security mechanisms and data encryption standards. A software-defined network (SDN) provides control over the normal network architecture to form user-controlled delivery services. The next generation of SDN needs to incorporate mobile networks or software-defined mobile networks (SDMNs) to facilitate the IoT. With the involvement of such a huge number of devices, security has become a major challenge in mobile-based SDN. Moreover, the need for SDMN with a 5G mobile network can generate huge mobile traffic in heterogeneous wireless environments.

This chapter discusses a four-layered IoT architecture for the smart city system. Cloud computing is an essential technology that is integrated into the IoT architecture to control the massive use of IoT devices and provide respective services to the end users. The four layers are the perception layer, network layer, service layer, and application layer. The perception layer consists of sensors, RFID, cameras, and other devices that are responsible for gathering information from the surroundings. These are integrated with real-world objects to provide smartness and connectivity. The network layer is responsible for connecting real-world objects to the internet. It handles the data transfer to the cloud through the basic wireless or wired network, local area network, and other networks. It incorporates various communication technologies such as Zigbee, Bluetooth, Wi-Fi, and data and cellular networks to communicate collected data to the service layer. This layer uses the virtualization technique whereby each physical object is digitally represented as a virtual object (VO) [23]. Each VO is implemented as a web service and can be related to a database to store the data sent by its physical counterparts. It enables interoperability among devices to communicate seamlessly irrespective of the communication technologies used. It helps in achieving a standardized communication platform between physical devices. Requests received from the upper layer to find services can be processed by the virtualization platform according to the requested service [24]. The service layer is responsible for the analysis of the service requested by the application in the layer above.

5.3 Analysis of Current Technology Available for Smart Cities

The design of a perfect smart city remains a long way off. Still, modern technological advances in IoT can form the basis of a smart and connected network of devices that can enhance our quality of life. Three of the areas that are directly responsible for the quality of service in IoT are the medium of connectivity, communication protocol, and storage of voluminous data.

The communication in IoT can use both wired and wireless communication mediums to keep the connectivity alive. The backbone of the internet is built around wired network standards. Most applications are made in terms of telephone, television and fiber-optic networks. Wired connections are more reliable and can transfer data at a relatively higher speed than wireless mediums. The future of IoT inevitably lies in wireless communication. However, the wired standard can still be useful for short-range communication. For example, devices in a vehicle can use wired communication to reduce the interference caused by passengers. Ethernet is the most popular standard for wired communication. Table 5.1 shows the ethernet standard available for networking. High-speed serial standards have also been in use in industry to increase communication efficiency between various components of smart devices. Table 5.2 shows some of the common standards available for serial communication. Serial communication needs no clock and is very space efficient. It also costs less and does not interfere with other communication devices. Similarly, many standards have been developed for wireless communication. The focus of such standards is to develop low-power, low-latency multi-way communication. Table 5.3 summarizes some of the standards popular for wireless communication.

TABLE 5.1

Ethernet Standards for Networking

Year	IEEE Std.	Name	Data Rate	Type of Cabling
1990	802.3i	10BASE-T	10 Mb/s	Category 3 cabling
1995	802.3U	100BASE-TX	100 Mb/s	Category 5 cabling
1998	802.3z	1000BASE-SX	1 Gb/s	Multimode fiber
1998	802.3z	1000BASE-LX/EX	1 Gb/s	Single-mode fiber
1999	802.3ab	1000BASE-T	1 Gb/s	Category 5e or higher category
2003	802.3ae	10GBASE-SR	10 Gb/s	Laser optimized
2003	802.3ae	10GBASE-LR/ER	10 Gb/s	MMF Single-mode fiber
2006	802.3an	10GBASE-T	10 Gb/s	Category 6A cabling
2010	802.3ba	40GBASE-SR4/LR4	40 Gb/s	Laser-optimized MMF or SMF
2010	802.3ba	100GBASE-SR10/ LR4/ER4	100 Gb/s	Laser-optimized MMF or SMF
2015	802.3bq	40GBASE-T	40 Gb/s	Category 8 (Class 1 & II) Cabling
2015	802.3bm	100GBASE-SR4	100 Gb/s	Laser-optimized MMF
2016	SG	Under development	400 Gb/s	Laser-optimized MMF or SMF

TABLE 5.2

Serial Communication Standards

Parameter	RS232	RS422	RS485
Cabling	Single ended	Differential	Differential
Number of devices	1 T–1 R	5 T–10 R	32 T–32 R
Communication mode	Full duplex	Full duplex/half-duplex	Half-duplex
Maximum distance	50 feet at 19.2 kbps	4000 feet at 100 kbps	4000 feet at 100 kbps
Maximum data rate	19.2 kbps at 50 feet	10 Mbps at 50 feet	10 Mbps at 50 feet
Signaling mode	Unbalanced	Balanced	Balanced
Current capability	500 mA	150 mA	250 mA

TABLE 5.3

Wireless Communication Standards

	Network Type	Year	Network Size	Bit Rate	Frequency	Range
802.11a	WLAN	1999	30	54 Mbps	5 GHz	120 m
802.11b	WLAN	1999	30	11 Mbps	2.4 GHz	140 m
802.11g	WLAN	2003	30	54 Mbps	2.4 GHz	140 m
802.11n	WLAN	2009	30	248 Mbps	2.4/5 GHz	50–250 m
802.11ac	WLAN	2014		3.2 Gbps	5 GHz	30 m
802.15.1	WPAN	2002/2005	7	3 Mbps	2.4 GHz	100 m
802.15.3	WPAN	2003	245	55 Mbps	2.4 GHz	100 m
802.15.4	WPAN	2007	65535	250 Kbps	868–915 MHz 2.4 GHz	75 m
802.15.6	WBAN	2011	250	10 Mbps	402–405 MHz	2–5 m

Short-range wireless communication standards like Bluetooth, Zigbee, Z-wave, LoRa, SigFox, etc., are designed to operate with low power and are designed specifically for IoT. Most IoT research works in a non-standard communication model. Every aspect of data sharing is standardized to make development faster. The process of standardization will also alleviate many of the security concerns related to open network communication. Similarly, the communication protocols need to address the lack of reliability in machine-to-machine communication.

As well as the communication model, the storage method can also impact the effectiveness of IoT. Local storage systems use NTFS, EXT4, HFS+, ZFS for storing large-volume data and providing a mechanism for data recovery. Similarly, network/distributed file systems like NFS, CIFS/Samba, AFP, Hadoop DFS, Ceph, etc., are also used. These provide data redundancy and scalability. However, there is still a need for a low-power, minimum-cost storage solution. IoT also needs storage hardware that can last a long time to increase the lifetime of deployed IoT networks. Researchers also need to focus on a modular approach to storage devices to reduce e-waste.

5.4 Overview of Power Consumption Models in IoT for Smart Cities

One of the major challenges in IoT-based smart cities is the amount of power required to sustain all devices to achieve ubiquity. To limit power consumption, intelligent connectivity, processing, and gathering algorithms are needed. First, however, the power consumption requirements for each component of the smart city need to be evaluated. Many power consumptions models have been proposed, in which total power consumption broadly depends upon sensing, processing, transmitting, and managing components [43]. The total power consumption can be expressed as follows.

$$P_t = P_s + P_p + P_r + P_m \tag{5.1}$$

where P_t is total power consumption, P_s is power required by sensors to collect the data from the environment, P_p is the power required to process the data to find useful information, P_r is the total power needed to transmit the consolidated information to destination over a connected network and P_m is the power used by the management system that manages the connectivity and provide service to the users.

Sensors are the most important part of IoT networks. The sensor converts real-world energy domain data to electrical domain data. Sensors can be active or passive according to the excitation signal needs. In smart cities most sensors are active sensors and need more power to collect environmental information. However, these sensing models generally waste a lot of power to maintain continuous data readings. Hence, the IoT model uses a low-power listening model to reduce the power [44]. Low-power listening turns sensors on and off periodically to avoid power wastage. The power consumption for a fixed period can be modeled as follows:

$$P_s = tP_{dc} + P_{sc} \tag{5.2}$$

where P_{dc} is the power per duty cycle with respect to type of sensor, t is time for which the sensor is on and collecting the environment data, and P_{sc} is the power consumed during the sleep period. Table 5.4 gives some examples of power consumption by different type of sensors.

The sensor nodes have microcontrollers that analyse data before transmitting it to the destination. The processor is the core of the calculation in nodes. The processor not only manipulates the data but also evaluates security threats and coordinates data transmission. The energy consumption in the processor depends on the load of the application and sleep time, i.e., major processor power consumption only occurs when it operates. The energy consumed for processing can be expressed as follows:

$$P_p = tV_{dc} * I_{dc} + P_{psc} \tag{5.3}$$

where t is the duty cycle, V_{dc} operating voltage, I_{dc} current drawn and P_{psc} is the sleep-time power consumption by the processor. Again, the power consumption in the processing

TABLE 5.4

Power Consumption by Various Types of Sensor Nodes

Sensor Type	Power Consumption (in mW)
Image sensor	150
Acceleration sensor	3
GPS receiver	15
PPG sensor	1.47
Humidity	1
3D accelerometer	0.32
Pressure sensor	15
Temperature sensor	0.5–5
Gas sensor	500–800

Note that the power consumption depends on the sensor manufacturer

TABLE 5.5

Power Drawn by Various Processing Units

Microcontroller	Active Mode (mW)	Sleep ADC (mW)	Sleep Idle (mW)	Sleep Standby (mW)
Arduino Uno	225	180	170	140
Arduino Micro	200	130	110	40
Arduino Nano	85	48	48	32

The power recorded here is with an operating voltage of 5V.

unit depends on the manufacturer and model of the unit [45]. Table 5.5 gives an overview of the power consumed by various versions of the microcontrollers.

In smart cities, data communication can be by a wired or wireless medium. In all cases the sensor will use power to transmit data from device to base station in a one-to-one manner. The power required for the base station, then, will depend upon the number of users accessing the station simultaneously. The exchange terminals (that are responsible for routing and locating destinations) will consume power [46]. This will also depend on the number of users accessing the terminal simultaneously. Finally, the destination node will consume power to receive the data. The overall power consumption per device per transmission can be summarized as follows:

$$P_r = 1.5\left(P_{sn} + \frac{P_{bn}}{N_{bn}} + \frac{P_{en}}{N_{en}} + P_{rn} \right) \tag{5.4}$$

where P_{sn} represents the power required by the sensing node to send the data to the base station, P_{bn} is the total power consumption of the base station during the transmission time, N_{bn} is the number of communications accessed the base station during the period, P_{en} is the power consumption by all of the exchange stations during communication, N_{en} is the number of other communications using the same exchange stations, and P_{rn} is the power required by the receiving node to maintain connection to the network. Finally, the multiplier of 1.5 is used to include other power consumption and losses.

The energy consumption of the sensor nodes usually depends upon the connection type used to connect the node to base station. In the case of wired connectivity, the power is usually a fixed value as power is lost after a very long transmission distance. However, in wireless the radio signal consumes more power based on distance and connectivity requirements. Moreover, the power consumption depends on the wireless technology used [46]. Hence, in the case of wireless node connection, the power consumption can be modeled as follows:

$$P_{sn}^w = P_{tx} + P_{rx} + \frac{P_{packet}}{L_{packet}} \tag{5.5}$$

where P_{tx} is the power required by transmitter to transmit a bit of data, P_{tx} is the receiver power consumption per bit, P_{packet} is the energy required for decoding a packet, and L_{packet} is the payload length in bits. The power analysis of various connection media[47] is summarized in Table 5.6.

Power consumption of the managing units includes the power consumed by routers, the central control unit and cloud servers. The accurate modeling of power needed by these units varies according to the distance from the source, the number of hops, and the

TABLE 5.6

Details of Various Connection Types in Wireless Medium

	Bluetooth	Zigbee	Wi-Fi
Awake mode	35 mA	50 mA	245 mA
Transmitting mode	39 mA	52 mA	251 mA
Receiving mode	37 mA	54 mA	248 mA
Power supply	3.3 v	3.3 v	5 V

size of data to be transferred. Precise conventional analysis of the power needed continues to challenge IoT researchers. Many have tried to address predictions by using neural network models [48, 49]. However, these models require improvement in order to deal with real-time usability.

5.5 Smart City Scenario

The service layer processes information received from the two layers in two events: service events and resource events. Service events control the services offered by the smart city, such as notifying the user of traffic congestion in a particular area [50, 51]. Resource events are responsible for managing resources such as water consumption, electricity consumption, and so on. The services are characterized using various factors such as information associated with the user, the output from the previous request, preferences, policies, and so on. A context-awareness mechanism is also followed to detect, recognize, classify, and act upon the situation the user is involved in. All this processing is done in the cloud servers. This layer uses various data management techniques such as big data, HDFS, and HBase to filter, analyze and process the data [25]. Further, it uses machine learning algorithms to classify the services for the respective applications in the smart city. The application layer is responsible for the final processing and output of the results to different user categories. It provides a user interface for exploiting the services of the smart city system.

Figure 5.1 represents the four-layer IoT architecture for the smart city system. The way the architecture works is described using the following real-world scenario. Data from roadside sensors keeping track of the number of vehicles are transferred to the service layer using communication technology such as data network, Zigbee, and cellular networks in the network layer. Network layers provide a virtualization platform between the sensor interface and the gateways. The service layer uses various data management tools such as MapReduce, HDFS, to filter, analyze, and process the data. Further, this information is classified according to the service event. These service events are shared with the corresponding applications. Event management is done according to the existing policies, information, and machine learning techniques. In this scenario, potential travelers can be notified about traffic congestion if a specific threshold is exceeded on the application interface. Subsequently, the application layer transfers control to the service layer services (in this case, traffic congestion service). The system can check for alternative routes by querying with the lower layers (perception and network)and can notify drivers on their application interface.

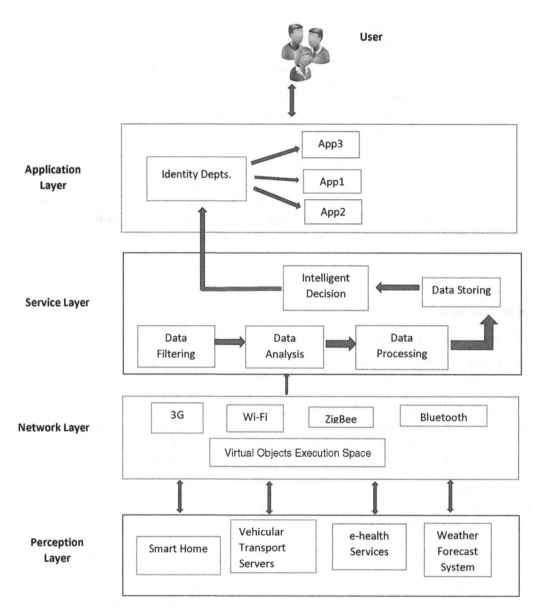

FIGURE 5.1
Layer IoT architecture of smart cities.

Apart from these four layers, the architecture includes a security engine that focuses on the implementation of the security procedure to provide reliable and secure communication at every level. It is responsible for implementing access control, handling authentication requests, and so on.

5.6 Smart City Aspects

This section considers aspects of smart cities such as smart business, smart citizen, smart home, smart education, smart healthcare, smart transportation, smart grid etc., which are illustrated in Figure 5.2.

5.6.1 Smart Citizen

Citizens are the core of smart cities. Personal data analysis provides numerous advantages, including individualized services, increased innovation, and economic growth [26, 27].

5.6.2 Smart Home

Many smart devices and items of equipment are found in a smart house. The smart home-owner can use a private blockchain to keep track of communication between local devices. A smart house of the future will have many smart devices. To provide some services, smart gadgets must communicate with one another. A light bulb, for example, must request data from the motion sensor to switch on the lights automatically when someone enters the

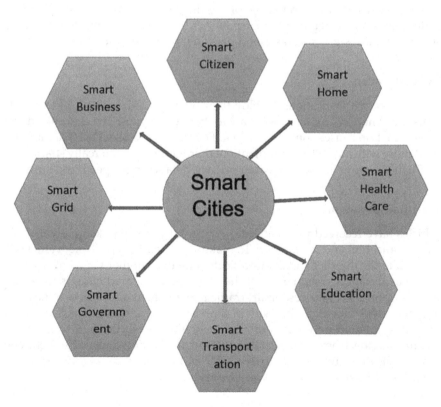

FIGURE 5.2
Aspects of smart cities.

house. Appliances in the home typically require energy. Prepaid cards or mobile devices that have been used in the past are frequently used as a method of payment for energy consumption. However, in the smart house of the future, automatic payment has become a trend that does not require human intervention [28, 29].

5.6.3 Smart Health Care

The foundation of a pleasant existence for citizens is good health. Advances in medical technology have benefited citizens significantly. However, as a result of growing urbanization, traditional healthcare is no longer sufficient for the majority of the world's population [30]. Inconsistency between ever-increasing demands combined with limited resources makes it difficult to address everyone's needs. The existing healthcare system has considerable difficulties in providing high-quality, low-cost care. These issues are exacerbated by an aging population, with an avalanche of chronic diseases and increased demand for healthcare services [31]. Furthermore, due to a lack of resources in some cities, obtaining good treatment is challenging. The current healthcare system must therefore change to become a smart healthcare system [32]. Intelligent healthcare requires wearable technology, smart hospitals, smart emergency response, and smart equipment. Patient data is critical to ensuring that patients receive the best possible care. Patient data sharing across hospitals can aid nurses and doctors in judging a patient's state and making real-time choices on patient health, even in remote places, in smart healthcare [34, 35]. In the smart health system, sensors are deployed in the patient's body, and different sensors can capture different physiological signs such as blood pressure, heart rate, etc.

5.6.4 Smart Education

Most higher education institutions employ their own specialized record-keeping systems for students' information, saving it using a variety of data types. The centralized storage of students' information causes some issues [33]. For example, a user can share a reference to his or her identity with any of the aforementioned programs without really sharing his or her identity with them, providing for increased privacy, data savings, speed, and authentication.

5.6.5 Smart Transportation

Smart vehicles have received much attention in recent years, thanks to advances in information communication and technology. To make smart vehicles possible, vehicles must be able to connect to the internet and communicate with one another via what is known as an intelligent transportation system. One of the main goals of smart cities is to provide new traffic and transportation services. Smart traffic management attempts to educate users by providing them with data that will result in safer and more informed decisions about how to use the transportation network. Different technologies can be used to improve traffic management, ranging from basic systems like car navigation and traffic-light control to monitoring applications that use security CCTV and motion sensors to integrate live data and feedback for advanced applications like parking.

5.6.6 Smart Government

Benefits of e-government include improving the quality of government services, developing individual credit systems, enhancing the government's legitimacy, and fostering resource integration. Electronic voting is another potential application of blockchain technology in the smart government domain [36, 37]. The challenges of government initiatives are information technology, organizational and managerial, environmental and institutional, etc.

5.6.7 Smart Grid

Smart grids have evolved a great deal since the conception of the first grid. Due to the inability to match generation to demand, traditional grids suffer from over-generation. Because it is difficult for distributors to accurately predict peak demand, they must over-produce energy to meet demand during peak hours, resulting in pollution and contributing to global warming where fossil fuels are used. Because of its distributed nature, the smart grid poses new security and privacy risks. These are dispersed throughout hundreds of square kilometers, with multiple substations that can each serve as a network access point. Physical security for such a dispersed system is extremely difficult to achieve [38, 39]. Renewable energy (e.g., solar and wind energy) is being used more broadly to safeguard the environment (e.g., atmosphere and water). Furthermore, with the development of new battery energy storage, a large number of consumers will become prosumers, able to generate and store electrical energy using renewable energy. A smart grid is recommended in this scenario to create an efficient, secure, economic, and sustainable power grid system [40, 41].

5.6.8 Smart Business

Smart city models must be secure and resilient to faults and dangers. These technologies enable the transformation of traditional homes and businesses into virtualized homes and businesses at a low cost. They reduce the number of hardware devices, as well as the cost of specialist maintenance and support. It is worth noting that smart city models minimize CAPEX while increasing OPEX as enterprises migrate to the cloud. As a result, in smart cities, strategies for lowering OPEX expenses are required [42].

5.7 Issues Related to Smart Cities

Due to the unique design requirements of smart cities, many challenges arise for the design and preservation of intelligent IoT systems. The system becomes more challenging due to the individual nature and requirements of each application. The following sections gives an insight into the challenges that may arise while designing various components of smart cities.

5.7.1 Human Well-being

Intelligent systems in smart cities take many decisions regarding human life, such as maintaining accident-free roads, alerting medical professionals health issues of patients, predicting threats, etc. Any small error can lead to potential loss of life or wealth. Hence, the design of smart applications must adhere to a critical evaluation process to provide a better service to human well-being.

5.7.2 Communication

Modern intelligent cities are designed to accommodate millions of devices and demand to establish a seamless connection between them. In such cases, connectivity becomes a major aspect of the design goal. The design process must include the process for the connection of masses of devices as well as alternatives for breakdown scenarios.

5.7.3 Transportation

Smart cities will have a densely connected network of roads. The management of logistics and traffic will be one of the major concerns for such a system. The traffic control system must be intelligent enough to be able to prioritize essential services and should be able to resolve unexpected congestion.

5.7.4 Power Efficiency

Applications, communication devices, and IoT devices in smart cities will consume a great deal of power to sustain connectivity and services. Cities should be designed in such a way as to use renewable power sources, and devices are to be designed to use minimal power to carry out their objectives. Design processes for other applications should waste as little power as possible.

5.7.5 Preservation of Critical Resources

Food and water will be another major concern for such massive cities. Hence, smart cities must include the design of non-traditional food production mechanisms like vertical farming and must manage water resources efficiently. Smart cities must also maintain intelligent logistics related to water and food supply.

5.7.6 Efficient Healthcare System

In densely populated smart cities, health issues will be a major concern. In a pandemic situation like COVID-19 the healthcare system must not fail. Effective communication and data collection will be key to making the city prepare for any health hazard situation. Moreover, the transportation and maintenance of essential medicines and life-saving equipment must be supervised intelligently. Prediction systems for any future health problems must be designed efficiently to provide pre-emptive solutions.

5.7.7 Security and Trust

As smart cities are to relate to a mesh of interconnected devices, security will be a major challenge. Smart cities applications use personal data to provide personalized services. Even a single data privacy breach can create panic. Strict guidelines related to data privacy must be followed by any applications used to increase trust in the system.

5.8 Usefulness and Challenges of the Smart City in India

India is one of the most densely populated countries in the world and has a large population living below the poverty line. Its cities became established hundreds of years before the advent of modern technology. Converting the existing city structure into a smart city is especially challenging in India. The state of the infrastructure available in these cities has declined over time. Hence, a perfect smart city is a dream for most of the population who are struggling with basic needs like water supply. Unpredicted governance is also one of the major obstacles in the development of smart cities in this country. The elected government not only have to deal with corruption, an overloaded judiciary, and constant threats from extremists, but also with limited revenue. The lack of transparency and coordination among bureaucrats further aggravates the situation. Conditions are improving gradually, and the development of smart cities is certainly rising in India. However, the sudden growth of cities and urbanization is causing sustainability problems. The sustainability of the economy, environment, and sovereignty is becoming a major challenge for the exponentially growing urban population. Some of the environmental impacts include poor air quality, lack of fresh water supply, heatwaves, artificial floods, and the pandemic. Likewise, poor returns on investment, a reduction in job opportunities, and lack of a skilled workforce are taking their toll on economic stability. The rise of extremist groups and unlawful activity is also hampering smart city development.

'Smart city' is currently being used as an advertising buzzword, with the only spotlight being on technological advances. However, the contribution of smart cities has to be sustainable and should aid human values. Drainage, sewerage, healthcare systems, transport, and education are some of the fields that can benefit from India's smart cities. The population of India needs to be educated on the impact of the environment on the current economy-based education system. Unity, rights and freedoms must be enforced by the judiciary over any political agenda. While these may not appear directly related to the development of smart cities, development can be directly affected by the governing body in India. Moreover, the message as to the reusability and repairability of products should be reinforced. Lastly, economic inequality needs to be reduced in order to develop cities that can assist in human growth.

The aim of smart cities should be about developing a coherent, sustainable society to build a smart nation. India is the most diverse and second-most populous country in the world and can benefit from smart city technologies. Smart technology will help to streamline the urban supply chain management to reduce wastage. It will enable better services for citizens in education, security, job opportunities, etc. Smart cities in India can take advantage of business intelligence to create jobs and improve the economic status of their citizens. As with smart cities, monitoring and actuation will become simple. Hence, efficient resource allocation and situation management will be easier to implement. It will also help to reduce greenhouse gas emissions by designing efficient traffic and reducing power requirements. Smart cities will make a major contribution in the education and economic fields, which will lead to a sustainable developed country.

5.9 Conclusion

In general, the major components of smart cities are smart citizen, smart home, smart grids, smart government, smart healthcare, smart education, and smart transportation, etc. Smart city development is hampered by a number of factors, including social, political, and, most importantly, technical challenges. The technological difficulties concern system compatibility and cost-effective technologies, necessitating a focus on security and privacy concerns. Because the networks that are a vital element of the operation of smart cities are vulnerable to hostile attackers, security is a top priority.

References

1. Davis, K. (1965). The urbanization of the human population. *Scientific American*, 213(3), 40–53.
2. 68% of the world population projected to live in urban areas by 2050, says UN, Department of Economic and Social Affairs, https://www.un.org/development/desa/en/news/population/2018-revision-of-world-urbanization-prospects.html, 2018.
3. Infographics: Urbanisation and Urban Development in India, URBANET, https://www.urbanet.info/urbanisation-in-india-infographics/, 2018.
4. Yang, X. J. (2013). China's rapid urbanization. *Science*, 342, 310.
5. What is Urbanization, Conserve Energy Future, https://www.conserve-energy-future.com/causes-effects-solutions-urbanization.php, 2021.
6. Gharaibeh, A., Salahuddin, M. A., Hussini, S. J., Khreishah, A., Khalil, I., Guizani, M., & Al-Fuqaha, A. (2017). Smart cities: A survey on data management, security, and enabling technologies. *IEEE Communications Surveys & Tutorials*, 19(4), 2456–2501.
7. Haughton, G. (1997). Developing sustainable urban development models. *Cities*, 14(4), 189–195.
8. Fernandez-Anez, V. (2016, June). Stakeholders approach to smart cities: A survey on smart city definitions. In *International Conference on Smart Cities* (pp. 157–167). Springer, Cham.
9. Songdo Project (n.d.) (last access: 12/09/2021) www.songdo.com
10. DesignBUILD (n.d.) (last access: 12/09/2021) http://designbuildsource.com.au/eco-city-in-china
11. Lavasa Highlights My City (n.d.) (last access: 12/09/2021) http://www.masdar.ae/en/#masdar
12. eSMARTCITY.es (n.d.) (last access: 12/09/2021) http://www.esmartcity.es/noticiasDetalle.aspx?id=6174&c=6&idm=10
13. Sheu, M. H., Chang, L. H., Hsia, S. C., & Sun, C. C. (2016, May). Intelligent system design for variable color temperature LED street light. In *2016 IEEE International Conference on Consumer Electronics-Taiwan (ICCE-TW)* (pp. 1–2). IEEE.
14. IBM. "IBMs smarter cities challenge: Boston report." 2012, (Accessed on December 2016). [Online]. Available: https://smartercitieschallenge.org/assets/cities/boston-unitedstates/documents/boston-united-states-full-report-2012.pdf
15. Zurborg, A. (2010). Unlocking customer value: the virtual power plant. *World Power*, 1–5.
16. Finch, K., & Tene, O. (2013). Welcome to the metropticon: Protecting privacy in a hyperconnected town. *Fordham Urb. LJ*, 41, 1581.
17. Mohanty, S. P., Choppali, U., & Kougianos, E. (2016). Everything you wanted to know about smart cities: The internet of things is the backbone. *IEEE Consumer Electronics Magazine*, 5(3), 60–70.
18. Maisonneuve, N., Stevens, M., Niessen, M. E., Hanappe, P., & Steels, L. (2009). Citizen noise pollution monitoring.
19. Li, X., Shu, W., Li, M., Huang, H. Y., Luo, P. E., & Wu, M. Y. (2008). Performance evaluation of vehicle-based mobile sensor networks for traffic monitoring. *IEEE transactions on vehicular technology*, 58(4), 1647–1653.
20. Nuortio, T., Kytöjoki, J., Niska, H., & Bräysy, O. (2006). Improved route planning and scheduling of waste collection and transport. *Expert Systems with Applications*, 30(2), 223–232.
21. Seike, H., Hamada, T., Sumitomo, T., & Koshizuka, N. (2018, October). Blockchain-based ubiquitous code ownership management system without hierarchical structure. In *2018 IEEE SmartWorld, Ubiquitous Intelligence & Computing, Advanced & Trusted Computing, Scalable Computing & Communications, Cloud & Big Data Computing, Internet of People and Smart City Innovation (SmartWorld/SCALCOM/UIC/ATC/CBDCom/IOP/SCI)* (pp. 271–276). IEEE.

22. Banerjee, M., Lee, J., & Choo, K. K. R. (2018). A blockchain future for internet of things security: a position paper. *Digital Communications and Networks*, *4*(3), 149–160.

23. Nitti, M., Pilloni, V., Giusto, D., & Popescu, V. (2017). IoT Architecture for a sustainable tourism application in a smart city environment. *Mobile Information Systems*, *2017*.

24. Zhang, C. (2020). Design and application of fog computing and Internet of Things service platform for smart city. *Future Generation Computer Systems*, *112*, 630–640.

25. Silva, B. N., Khan, M., & Han, K. (2020). Integration of big data analytics embedded smart city architecture with RESTful web of things for efficient service provision and energy management. *Future Generation Computer Systems*, *107*, 975–987.

26. Lai, A., Zhang, C., & Busovaca, S. (2013). 2-square: A web-based enhancement of square privacy and security requirements engineering. *International Journal of Software Innovation (IJSI)*, *1*(1), 41–53.

27. Armbrust, M., Fox, A., Griffith, R., Joseph, A. D., Katz, R., Konwinski, A., ... & Zaharia, M. (2010). A view of cloud computing. *Communications of the ACM*, *53*(4), 50–58.

28. Dorri, A., Kanhere, S. S., Jurdak, R., & Gauravaram, P. (2017, March). Blockchain for IoT security and privacy: The case study of a smart home. In *2017 IEEE International Conference on Pervasive Computing and Communications Workshops (PerCom Workshops)* (pp. 618–623). IEEE.

29. Xu, A., Li, M., Huang, X., Xue, N., Zhang, J., & Sheng, Q. (2016). A blockchain based micro payment system for smart devices. *Signature*, *256*(4936), 115.

30. Collins, F. S. (2015). Exceptional opportunities in medical science: a view from the National Institutes of Health. *JAMA*, *313*(2), 131–132.

31. Acampora, G., Cook, D. J., Rashidi, P., & Vasilakos, A. V. (2013). A survey on ambient intelligence in healthcare. *Proceedings of the IEEE*, *101*(12), 2470–2494.

32. Gharaibeh, A., Salahuddin, M. A., Hussini, S. J., Khreishah, A., Khalil, I., Guizani, M., & Al-Fuqaha, A. (2017). Smart cities: A survey on data management, security, and enabling technologies. *IEEE Communications Surveys & Tutorials*, *19*(4), 2456–2501.

33. Turkanović, M., Hölbl, M., Košič, K., Heričko, M., & Kamišalić, A. (2018). EduCTX: A blockchain-based higher education credit platform. *IEEE Access*, *6*, 5112–5127.

34. Kuo, T. T., Kim, H. E., & Ohno-Machado, L. (2017). Blockchain distributed ledger technologies for biomedical and health care applications. *Journal of the American Medical Informatics Association*, *24*(6), 1211–1220.

35. Mettler, M. (2016, September). Blockchain technology in healthcare: The revolution starts here. In *2016 IEEE 18th International Conference on e-Health Networking, Applications and Services (Healthcom)* (pp. 1–3). IEEE.

36. Yavuz, E., Koç, A. K., Çabuk, U. C., & Dalkılıç, G. (2018, March). Towards secure e-voting using ethereum blockchain. In *2018 6th International Symposium on Digital Forensic and Security (ISDFS)* (pp. 1–7). IEEE.

37. Zhao, J. J., & Zhao, S. Y. (2010). Opportunities and threats: A security assessment of state e-government websites. *Government Information Quarterly*, *27*(1), 49–56.

38. Mirian, A., Ma, Z., Adrian, D., Tischer, M., Chuenchujit, T., Yardley, T., ... & Bailey, M. (2016, December). An internet-wide view of ICS devices. In *2016 14th Annual Conference on Privacy, Security and Trust (PST)* (pp. 96–103). IEEE.

39. Formby, D., Srinivasan, P., Leonard, A. M., Rogers, J. D., & Beyah, R. A. (2016, February). Who's in Control of Your Control System? Device Fingerprinting for Cyber-Physical Systems. In *NDSS*.

40. Peck, M. E., & Wagman, D. (2017). Energy trading for fun and profit buy your neighbor's rooftop solar power or sell your own-it'll all be on a blockchain. *IEEE Spectrum*, *54*(10), 56–61.

41. Guerard, G., Pichon, B., & Nehai, Z. (2017, April). Demand-Response: Let the Devices Take our Decisions. In *SMARTGREENS* (pp. 119–126).

42. Yang, B., Kim, Y., & Yoo, C. (2013). The integrated mobile advertising model: The effects of technology-and emotion-based evaluations. *Journal of Business Research*, *66*(9), 1345–1352.

43. Martinez, B., Monton, M., Vilajosana, I., & Prades, J. D. (2015). The power of models: Modeling power consumption for IoT devices. *IEEE Sensors Journal*, *15*(10), 5777–5789.

44. Sha, M., Hackmann, G., & Lu, C. (2013, April). Energy-efficient low power listening for wireless sensor networks in noisy environments. In *Proceedings of the 12th International Conference on Information Processing in Sensor Networks* (pp. 277–288).

45. Widhalm, D., Goeschka, K. M., & Kastner, W. (2021, October). Is Arduino a suitable platform for sensor nodes?. In *IECON 2021–47th Annual Conference of the IEEE Industrial Electronics Society* (pp. 1–6). IEEE.

46. Haapola, J., Shelby, Z., Pomalaza-Raez, C. A., & Mähönen, P. (2005, February).Cross-layer energy analysis of multihop wireless sensor networks. In *EWSN* (Vol. 5, pp. 33–44).

47. Dementyev, A., Hodges, S., Taylor, S., & Smith, J. (2013, April). Power consumption analysis of Bluetooth Low Energy, ZigBee and ANT sensor nodes in a cyclic sleep scenario. In *2013 IEEE International Wireless Symposium (IWS)* (pp. 1–4). IEEE.

48. Alenazi, M. M., Yosuf, B. A., El-Gorashi, T., & Elmirghani, J. M. (2020, July). Energy efficient neural network embedding in IoT over passive optical networks. In *2020 22nd International Conference on Transparent Optical Networks (ICTON)* (pp. 1–6). IEEE.

49. Sahoo, K. S., Tiwary, M., Luhach, A. K., Nayyar, A., Choo, K. K. R., & Bilal, M. (2021). Demand-supply based economic model for resource provisioning in industrial IoT traffic. *IEEE Internet of Things Journal.*

50. Maity, P., Saxena, S., Srivastava, S., Sahoo, K. S., Pradhan, A. K., & Kumar, N. (2021). An effective probabilistic technique for DDoS detection in OpenFlow controller. *IEEE Systems Journal.*

51. Neha, B., Panda, S. K., Sahu, P. K., Sahoo, K. S., & Gandomi, A. H. (2022). A systematic review on osmotic computing. *ACM Transactions on Internet of Things*, 3(2), 1–30.

6

Performance Evaluation Methods for SDN Controllers

A Comparative Analysis

Jehad Ali and Byeong-hee Roh
Ajou University, South Korea

Sahib Khan
University of Engineering and Technology, Mardan

CONTENTS

DOI: 10.1201/9781003213871-6

6.1 Introduction

Due to rapid advances in internet technology, network terminals have become all but ubiquitous. Legacy network design, on the other hand, has failed to take future communication and internet technology improvements into account. Legacy network infrastructure could not keep up with the rapid development of the internet because it was out of date. Data and control planes are closely connected in conventional network design, which has various constraints. For example, if we wish to modify the configuration of the network, each device must be configured individually throughout the whole network. This is a significant disadvantage. For the same reason, suppliers do not provide developers and members of the community with access to the essential configuration settings of their devices, since doing so might cause networks to malfunction. Protocols are also deeply ingrained in the network equipment's firmware. Due to proprietary hardware and a lack of testing for creative networking solutions, these limitations limit network innovation, increase administrative effort, and drive up network management costs [1–4].

The SDN paradigm [5–8] introduces novelty in computer networks by decoupling the data and control planes. The separation of data and control planes shifts the network complexity from networking devices to the intelligent SDN controllers; thus, the network devices can be programmed through applications running on the controller, while the network is abstracted from applications [9] running on the top of the controller. The network functions according to the applications being executed on the SDN controller. As a result, the complexity of the networks reduces as the logic shifts from the devices to the centralized SDN controller.

However, besides the numerous advantages of SDN, its modeling, evaluation, and testing present several challenges. One of the challenges is the performance evaluation of the controllers [10–12]. The controller plays a prominent role in SDN. Hence, different approaches are used for performance evaluation and comparison of SDN controllers. However, these approaches need proper categorization to enable improvements in SDN research. This chapter discusses, categorizes and analyzes these approaches.

The objectives of the chapter are as follows:

- To explore controller performance evaluation procedures,
- To investigate several schemes for performance evaluation of SDN controllers based upon quantitative and qualitative analysis,
- To evaluate quantitative performance evaluation approaches in Mininet and Cbench,
- To evaluate feature-based comparison of controllers, and
- To evaluate hybrid methods combining quantitative performance evaluation and feature-based analysis.

Organization of the Chapter

The rest of the chapter is organized as: Section 6.2 discuss the basics of the SDN architecture and the important features in SDN controller performance evaluation. Section 6.3 explains various schemes in the literature for controller performance evaluation. Section 6.4

describes the analytical network process for controller selection in SDN. Section 6.5 illustrates the results. Section 6.6 concludes the chapter with future scope.

6.2 SDN Architecture

Generally, the SDN is composed of data, control, and management planes. In this section, we discuss each plane briefly. Figure 6.1 shows these three planes and the interaction among them through southbound (SB) and northbound (NB) APIs.

6.2.1 Data Plane

In SDN, the data plane comprises forwarding devices (known as the infrastructure or the underlying network). The data plane packets are matched and actions are taken in accordance with the forwarding rules described in a flow table.

A flow table is made up of many flow entries. The packet header information is compared to the flow table entries. Each flow entry consists of three fields: a header, an action, and a counter. Table 6.1 illustrates a flow table, with the top row containing header data and the subsequent rows containing flow entries. When a new packet arrives on a switch's ingress port, the matching procedure begins, as shown in Table 6.1. If the packet's destination IP address begins with 192.168.X.X, it is forwarded to port number 1, and counter 102 is updated. Similarly, the third row specifies that if a packet's source and destination port

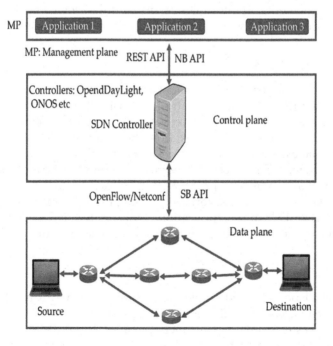

FIGURE 6.1
Three-plane architecture for SDN [9].

TABLE 6.1

An Example of the OpenFlow Table

Source IP	Destination IP	Source Port	Destination Port	Action	Counter
X	192.168.X.X	X	X	Port 1	102
X	X	21	21	Drop	621
192.168.1.X	X	X	X	Port 3	100
X	X	20	80	Port 4	101

numbers are identical, it should be dropped. If no rules for the new packet exist in the flow table, the switch sends a packet-in message to the controller, which returns the destination to the forwarding device (packet-out message) and updates the flow rules in the flow table accordingly. This is in contrast to typical networks, where routing decisions are made between closely coupled devices.

6.2.1.1 Southbound (SB) APIs

The SB-API interfaces the data and control planes. Various protocols, such as OpenFlow [13] and Netconf [14], are available for interfacing the two planes. OpenFlow is the most widely used protocol in the SDN community. OpenFlow provides a secure communication channel between the controller and the switches. The white paper [15] discusses the benefits and programmability of OpenFlow for forwarding device programming. The OpenFlow idea originated at Stanford University, and the Open Networking Foundation (ONF) [16] consortium currently manages the OpenFlow standardization process.

6.2.1.2 Northbound (NB) API

The NB-API interfaces the control and management planes. By employing the representational state transfer (REST) API, the controller acts as a bridge between the forwarding devices and the management plane. Similarly, this API is used to retrieve operational statistics (for example, regarding flow entries) from the data plane. Applications operating in the management plane connect with the controller using this API, and the data plane executes the relevant actions. These actions determine the behavior of the management plane application. For instance, a firewall program establishes rules that regulate the admission and egress of packets across the network. As a result, the data plane devices will forward or restrict traffic according to the application's policies. Similarly, a load-balancing program will manage traffic by detecting congestion in various network channels.

6.2.2 Management Plane

The management plane is located above the control plane, and it is where several programs may be run to accomplish the diverse activities required for an effective network. The data plane makes use of the management plane's flexibility and programmability, as well as the abstractions offered by the control plane. For instance, network monitoring may be accomplished comprehensively by the development and deployment of a snipping program. Other security programs may be used to identify and mitigate distributed denial of service (DDoS) threats in a similar fashion [17].

6.2.3 Control Plane

The control plane constitutes the most prominent part of the SDN infrastructure because the entire network is dependent upon it. It is implemented through the SDN controller. Several SDN controllers are available, the most important of which are NOX [18], POX [19], Ryu [20], Floodlight [21], TREMA [22], ODL [23] and ONOS [24]. Each controller has a number of features as discussed below.

(1) *OpenFlow:* OpenFlow is also known as the SB-API which directs the flow requests forwarded by the switches to the controller and vice versa. The controller sends flow response (PACKET_OUT) messages in response to flow request (PACKET_IN) messages. Thus, OpenFlow [13] manages the exchange of messages between these two planes. The effect on SDN controller delay of the exchange of request and response packets is discussed in [25]. However, different versions of OpenFlow are supported by each controller, for example, v1.3 has support for load balancing that facilitates performance improvement during high traffic generation by network devices.

(2) *GUI:* A graphical user interface (GUI) for a controller helps in obtaining statistical information about the underlying forwarding devices, the configuration of switches' flow entries and application deployment. The GUI is considered an important feature in controller selection for SDN and in qualitative and experimental analysis of the controllers [26]. Generally, the SDN controllers are configured through a command-line interface (CLI) or a GUI, or both. The statistics provided by the GUI of the controller consist of information regarding hosts and forwarding devices, flow entries, flow tables and creation of the SDN topologies [27]. Viewing these statistics through the GUI is user friendly and they can be easily examined for analysis. Likewise, flow entries can be inserted to the switches via the GUI, though its features affect performance because the execution of the GUI is slower than the CLI. Further, the GUI support with SDN controllers can be divided into two categories: web-based supported by Java, and Python based. The execution speed of controllers coded in Python is slow because of less support for fast memory access and multithreading.

(3) *Northbound REST API:* In the management plane, communication between applications and controller is through a REST API. Similarly, operational statistics for switches and topologies are collected via this API. The controller uses the REST API and acts as a bridge between the management plane and the data plane. This is a significant aspect of controller selection since it communicates with the controller directly. Rapid response from the API reduces latency and improves throughput. Hence, the REST API is important for SDN performance [28, 29] and controller selection.

(4) *Clustering:* Native support for clustering tends to improve the scalability, stability, and performance of the controller. Controllers that support clustering have improved performance with respect to latency [30]. Similarly, decreased latency is observed with increased switches. The same effect is observed during high traffic generation through the clustering of controllers.

(5) *Quantum API:* Users leverage the services provided by the cloud by calling them remotely over the internet. Thus, there is a competition between cloud service providers (CSPs) and network service providers (NSPs). Huang et al. [31] proposed an economic model that distinguishes the competition between CSP and NSP.

Similarly, the research conducted by Huang et al. in [32] described a way to provide E2E performance using a cloud services configuration model. With support for the Quantum API, SDN controllers can take advantage of cloud computing. Controllers that natively support this API can use the Quantum API to leverage the cloud in the realm of high-performance computing as well as OpenStack networking. Controllers with quantum support feature parallelism and fast memory access. Consequently, performance improves with SDN scalability, i.e., when the number of switches increases. This API was incorporated into a research study describing controller selection [28].

(6) *Synchronization:* This shows how effectively a controller stores and responds to information about the network, which affects the time taken to discover the topology. This is very important for performance monitoring in SDN [33]; controllers which take a short time to discover the topology improve SDN performance.

(7) *Productivity:* This relates to the ease of development of applications and to the programming language of the controller. Using a controller coded in Python makes it easy to develop your application, but without the platform support and multithreading, it is slow. Productivity is therefore, inversely related to controller performance [34]. Well-coded Python is easier to develop applications and more productive, but lower productivity means better Java support. This is due to the fast memory access, cross-platform support, multithreading, and inter-process communication (IPC) features of Java-supported controllers which lead to high performance.

(8) *Partnership support:* Many local and international organizations support SDN controllers. As a result, IT organizations not only consider technical strengths when choosing a controller but also key aspects such as the financial resources as well as the sponsors associated with a controller. This function plays a vital role in controller selection [34].

(9) *Platform support:* This concerns the compatibility of SDN controllers with different operating systems, such as Linux, Mac, or Windows. Executing compatibility across various platforms means the controller has support for multithreading, flexible management of memory and fast memory access affecting controller performance. Because controllers can create clusters on different platforms, running on other platforms improves the efficiency of clustering. This reduces latency and improves QoS during normal and high traffic loads [30].

(10) *Modularity:* The ability to make the sub-routines of the main program is called modularity. Modular support helps the controller in dealing with large-scale systems. If controller modularity is high, it means the sub-modules have the capability to run in parallel, which speeds up execution and hence reduces response time. It helps to improve performance, especially when scalability increases.

6.3 Categorization of SDN Controller Selection Approaches

Several methods of SDN controller selection are described in the literature. These methods can be divided into three main categories: comparing controllers by performance, features, and hybrid. The hybrid approach combines feature results with performance-based comparisons to select the best controller. These methods are described below.

6.3.1 Performance Analysis Controller Selection

The studies presented in [35], [36, 37] and [38, 39] compare the controllers through performance. The performance-based approach only considers the performance and neglects features of the SDN controller. This approach only considers the performance of a typical topology made in Mininet or through creating virtual hosts and switches in Cbench. Hence, realistic scenarios of the real internet are not taken into consideration in the experiment.

6.3.2 Feature-Based Controller Selection

The research carried out in [40–43] compares only the supporting features of SDN controllers. Examples of support features are platform, OpenFlow, REST API, and clustering. Another disadvantage of this approach is that it provides only a theoretical analysis of the feature set that the controller provides. This selection method will result in a cognitive overload. Thus, the optimal decision cannot be made due to the 7 ± 2 problem, or the limitations of the human memory for processing information, also known as Miler's law [44].

6.3.3 Hybrid Methods of Controller Selection

A comparison of four SDN controllers using a hybrid approach was highlighted by Bispo et al. [26]. The authors chose two controllers based on heuristic decisions in the table of nine controller functions and ran Cbench in throughput and latency mode to evaluate the performance of those controllers. The study did not provide an explicit ranking of these controllers as study only analyzed the feature table, making accurate choices impossible. Second, authors did not consider performance comparisons in a real internet topology. The study by Salman et al. [34] compared the six SDN controllers Opendaylight (ODL), Beacon [45], Nox, Maestro [46], Libfluid Raw [47] and Ryu with respect to an increase in the number of switches and threads in the latency and throughput modes. Another study by Mamushiane [48] compared the performance of four controllers consisting of Ryu, ODL, Open Network Operating System (ONOS), and Floodlight. In this study, the authors used Cbench for measuring latency and throughput. Anderson [33] compared five SDN controllers – Trema [22], Ryu, ODL, ONOS and Floodlight – for wireless networks using qualitative and quantitative studies. A qualitative study of two features of these controllers – clustering support and state handling – was carried out first. Thus, in case of switch or controller failure, information about the state was tabulated for the five controllers to see how each controller collects and stores network state information and the status of this information, i.e., whether the controller reloads this information from a previously saved state or relearns the network state. Likewise, the clustering information was tabulated to see if these controllers support clustering and how other controllers share information among the clusters of controllers. In the study, two controllers were selected based on these two features that met the constraints of the aerial networks. The performance of the two controllers was then evaluated through emulated experimental scenarios in Mininet. Nevertheless, the selection process for controllers is based on heuristic decisions, and as the number of controllers and features expands, cognitive overload can occur, leading to disobedience of Miller's law.

6.3.4 Multi-Criteria Decision-Making (MCDM) Methods for Controller Selection in SDN

MCDM is a mathematical decision-making technique used for the selection of alternatives based on certain criteria elements [49]. It has been extensively used in many fields, such as for strategy selection in software development [50], for natural resources management

[51] and for selection of a network among heterogeneous networks [52], etc. Various approaches are in use for the selection process, which depends on several criteria elements to achieve the desired objective, for example, TOPSIS, analytical hierarchy process (AHP) and ANP, etc. An AHP method of selecting an SDN controller was proposed by Khondoker et al. [28]. The research sought to select a controller from ten candidates considering only their features, but did not perform any experimental evaluation of the controllers, nor did it provide the mathematical details of the approach.

A hybrid scheme based on AHP which utilizes both the features and performance evaluation for controller selection is described in [29]. Based on the feature set selection, the three highest-ranked controllers were evaluated using Cbench, but the performance in real internet topologies was not evaluated. AHP does not take into account the feedback from the alternatives, and it only considers the criteria weights for selection of alternatives. Another problem with AHP is that it does not consider the dependency of the criteria elements, making precise and accurate selection impossible. In [53], ANP was used to model risk factors in large projects using risk indices. Similarly, Nazir et al. [54] used quality criteria to select software components. ANP has been used in wireless sensors for optimal cluster head selection [55]. Therefore, it is assumed that the ANP approach may be used to analyze systems with complex behaviors and structures. Due to the complexity of the systems, the dependencies between them have increased. Therefore, research on interdependent network systems is important [56]. ANP is an established tool for the decision-making process where dependency exists among criteria elements.

6.4 Analytical Network Process-Based Controller Selection

The ANP MCDM problem is formulated by first defining the aim or objective, then defining the parameters for the criteria or sub-criteria, and then considering the assessment options. After listing the controller's characteristics, ANP is applied. Figure 6.2 illustrates the technique in detail. The purpose of this research is to determine the optimal SDN controller based on the ten characteristics listed in Table 6.2. Equations (6.1) and (6.2) indicate the criteria and alternatives. F denotes the available features of the various SDN controllers, while C denotes the alternatives. It is necessary to create a network model that represents the criteria and options, as well as their connection. Each possibility was analyzed in terms of each criterion and vice versa using the network model.

$$F = \left(F_1, F_2, F_3, \ldots, F_N \right) \tag{6.1}$$

$$C = \left(C_1, C_2, C_3, \ldots, C_N \right) \tag{6.2}$$

The ten critical characteristics that should be examined while selecting SDN controllers are described below. We presume that each of these characteristics is necessary for the controller selection process. However, since controllers are constantly changing, we use the most up-to-date information on these aspects from controller documentation and research reported in [25, 26]. These critical characteristics are taken into account in the ANP-based optimal controller selection procedure. The significance of a feature in a controller is

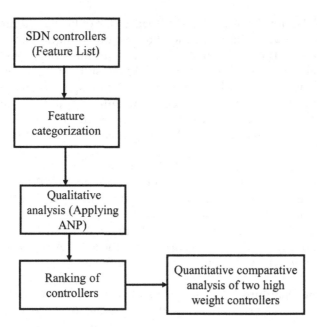

FIGURE 6.2
The procedure for controller selection leveraging ANP.

TABLE 6.2

Features Categorization for SDN Controller Selection

Alternatives	Criteria/Features									
	F_1	F_2	F_3	F_4	F_5	F_6	F_7	F_8	F_9	F_{10}
C_1	M	VH	Yes	Yes	No	M	M	L	L	M
C_2	H	H	Yes	Yes	Yes	M	M	VH	H	H
C_3	H	H	Yes	Yes	Yes	L	M	H	H	H
C_4	L	M	No	No	No	L	H	L	H	L
C_5	VH	L	No	Yes	No	H	H	M	L	M
C_6	L	L	No	Yes	No	L	H	L	L	M

determined by classifying these characteristics. A controller may handle two distinct sorts of features: ordinal and regular. Ordinal features have an intrinsic ordering, but regular categorical features do not. The classification of the feature set shows the extent to which each feature is supported by each controller. For example, C_4 and C_6 support only OpenFlow v1.0, and hence fall into the low category (L) for this feature (F1). C_1 support is medium (M), C_2 and C_3 support v1.0,1.1,1.3, respectively, and are therefore classified as high (H), whereas C_5 supports higher versions of OpenFlow, namely 1.5, and is thus classified as very high (VH). F_2 denotes a controller's GUI. C_1 supports a Java-based web interface and is quicker to run due to the presence of graphical tools for application and data plane administration.

As a result, F_2 is classified as very high due to its usage of Java multithreading. Similarly, C_2 and C_3 feature a Java-based interface and allow for the configuration of QoS parameters for data plane devices. They offer more application administration and topology setup options, which results in a slower GUI than C1. As a result, they are classified as high. C_4 has a Python-based interface and has a quicker execution speed than C_5 and C_6 due to its preparatory functions; hence, it is classified as medium. C_5 and C_6 offer solely Python-based interfaces; nevertheless, they perform poorly owing to the increased number of functions to maintain the data plane and management plane, as well as the absence of multithreading.

F_3, F_4, and F_5 are regular categorical traits, i.e., characteristics that cannot be subdivided further. A controller, for example, may or may not support REST APIs, open stack networking, or clustering. As a result, these features (F_3, F_4, and F_5) lack intrinsic ordering. In Table 6.2, these characteristics are denoted by Yes or No. As with C_4, C_5 and C_6 lack built-in support for REST APIs; hence, they get a No in the equivalent F_3 column of Table 6.2. Similarly, C_1, C_2, and C_3 provide built-in support for the REST API; hence, the F_3 column is marked Yes. F_4 demonstrates the Quantum API. Because C_1, C_2, C_3, C_5, and C_6 have an inbuilt support for the Quantum API, a value of Yes is shown in column F_4 for them. C_4 does not support the Quantum API; hence, No appears in F_4 for this controller. The clustering characteristic is denoted by F_5. Because the controllers C_1, C_4, C_5, and C_6 lack built-in capability for clustering, a No is entered in the F_5 column of Table 6.2. In comparison, C_2 and C_3 support clustering and are therefore denoted by Yes.

F_6 denotes the synchronization characteristic that has an effect on the data plane's topology discovery and response. C_1, C_5, and C_6 have a medium amount of interaction, which means that their interactions are rather sluggish. C_2 and C_3 are considered high level due to their rapid discovery of the underlying topology, whilst C_4 is considered low level due to the slowest interaction between the data and control planes. F_7 indicates a controller's production level. Productivity is connected to the simplicity with which an application may be developed and is dependent on the programming language used to write the controller. While developing applications using Python-coded controllers is straightforward, their lack of platform support, memory management, and multithreading make them sluggish. C_1, C_2, and C_3 have a moderate degree of productivity, whilst C_4, C_5, and C_6 have a high amount.

F_8 indicates the presence of support from many manufacturers. C_2 is backed by Cisco, NEC, IBM, and the Linux Foundation, which has a membership of over 40 corporations; as a result, it is rated very high. C_3 is backed by SK Telecom (South Korean telecommunications), Cisco, and NEC, and is therefore classified as high. C_1, C_4, and C_6 are supported by Big Switch Networks, Nicira, and NEC, respectively, and are therefore classified as low in terms of support. Although it is not directly connected to performance, effective vendor support ultimately results in performance improvement. F_9 is the platform support key. C_2, C_3, and C_4 all support three platforms, namely Linux, Mac, and Windows, and are therefore classified as high end. However, since C_1, C_5, and C_6 are supported on just one platform, namely Linux, they are accorded a low grade. Cross-platform compatibility allows multithreading and clustering, resulting in an increased quality of service. F_{10} indicates that modularity is supported. C_1, C_5, and C_6 have a moderate amount of modularity, but C_2 and C_3 have a high level of modularity. This is because C_2 and C_3 controllers may invoke sub-modules from the main function, resulting in parallel processing and therefore increasing performance. Categorization of features is performed as a pre-processing step prior to creating the comparison matrix.

6.4.1 Pairwise Comparison Matrix for Criteria and Alternatives

The pairwise comparison matrix is drawn up according to the 9-point scale proposed by Saaty [57] as shown in Table 6.3. It shows the relative importance of different components (criteria or alternatives) regarding an element. The same matrix is also employed to extrapolate the effect of the components on the objective using the 9-point scale as shown in Table 6.3. The values in the matrix are assigned to the criteria as well as alternatives representing personal judgments. For example, the importance of C_1 compared with C_2, C_3, C_4, and C_5 with respect to the F_1 component is assigned a value from the scale table. The value of $a_{(i,j)}$ represents the relative significance of a component corresponding to the ith row and row j^{th} column. The value of $a_{(i,j)} = 1$ in the pairwise comparison matrix shows the equal importance of the component corresponding to the ith row and j^{th} column. The diagonal components correspond to the comparison of the same components; therefore, their values are 1. The values below the diagonal are the reciprocal of the values above the diagonal. The value of, $a_{(5,1)} = 6$ shows that component in the 5th row is significantly to remarkably more important than the component in the 1st column. The value of $a_{(1,5)} = \dfrac{1}{a_{(5,1)}} = \dfrac{1}{6}$ is the reciprocal of $a_{(5,1)}$, denoting that a component in the 1st row is significant to remarkably less important than a component in the 5th column. The values are incorporated prudently for all the components in the pairwise comparison matrix.

6.4.2 Pairwise Comparison Matrix for Criteria and Alternatives

Alternatives are pairwise compared to each criteria component. The general form of the pairwise comparison matrix is denoted in the matrix (6.3). The rows and columns of the matrix are represented as M_1 to M_n and N_1 to N_n. First, the alternatives are pairwise compared with respect to the F_1 criterion. The values have been incorporated in (6.3) based using the 9-point scale defined in Table 6.3. The nonreciprocal and reciprocal values indicate the relative importance of the row and column components, respectively. First, the comparison of C_1 is made with C_2, C_3, C_4, C_5, and C_6 considering F_1 criterion. C_1 is of the same importance as itself therefore $a_{(1,1)}=1$. Then C_2 and C_3 are moderately more important

TABLE 6.3

Values for Relative Importance of One Feature/Controller Over Another

Scale	Description
1	Equally important
2	Equally to moderately more important
3	Moderately more important
4	Moderately to significantly more important
5	Significantly more important
6	Significantly to remarkably more important
7	Remarkably more important
8	Remarkably to excessively more important
9	Excessively more important

than C_1. i.e., $a_{(1,2)} = a_{(1,3)} = \dfrac{1}{3}$. C_1 moderately more important than C_4 and C_6, e.g., $a_{(1,6)} = 3$ shows that the alternative in this row (C_1) is moderately more important than the alternative in the corresponding column (C_6). $a_{(1,5)} = \dfrac{1}{6}$ shows that C_5 is significant to remarkably more important than C_1. Similarly, the values are filled for C_2–C_6.

$$
\begin{bmatrix}
 & N_1 & N_2 & N_3 & \cdots & N_n \\
M_1 & 1 & a_{(1,2)} & a_{(1,3)} & \cdots & a_{(1,n)} \\
M_2 & \dfrac{1}{a_{(1,2)}} & 1 & a_{(2,3)} & \cdots & a_{(2,n)} \\
M_3 & \dfrac{1}{a_{(1,3)}} & \dfrac{1}{a_{(2,3)}} & 1 & \cdots & a_{(3,n)} \\
\vdots & \vdots & \vdots & \vdots & \ddots & \vdots \\
M_n & \dfrac{1}{a_{(1,n)}} & \dfrac{1}{a_{(2,n)}} & \dfrac{1}{a_{(3,n)}} & \cdots & 1
\end{bmatrix}
\tag{6.3}
$$

Then, matrix (6.4) is used for summing up and each value is divided by the sum of the total values of the column. The next step is to find the eigenvector from the normalized matrix. The eigenvector shows the priority of these features F_1.

$$
\begin{bmatrix}
\dfrac{a_{(1,1)}}{\sum_{i=1}^{n} a_{(i,1)}} & \cdots & \dfrac{a_{(1,n)}}{\sum_{i=1}^{n} a_{(i,n)}} \\
\vdots & \ddots & \vdots \\
\dfrac{a_{(n,1)}}{\sum_{i=1}^{n} a_{(i,1)}} & \cdots & \dfrac{a_{(n,n)}}{\sum_{i=1}^{n} a_{(i,n)}}
\end{bmatrix}
\tag{6.4}
$$

The eigenvector X is obtained from the normalized matrix (6.4) according to Equation (6.5).

$$
X_i = \frac{1}{n} \sum_{j=1}^{n} a_{(i,j)}, \text{where } i = 1, 2, 3, \ldots, n
\tag{6.5}
$$

The result from Equation (6.5) is considered as the eigenvector X_1. To verify whether the judgments made while making the pairwise matrix are consistent, the next step is to find the CI and CR values. However, before making the consistency analysis, the consistency measure *(CM)* vector is to be calculated.

Consistency Measure: The *CM* vector is a prerequisite for the calculation of CI and CR. The consistency measure is calculated according to Equation (6.6). M_j denotes the row of the comparison matrix (6.3). X and x_i represents the eigenvector and the corresponding element of the eigenvector as shown in the matrix (6.5). The M_j and X are multiplied and

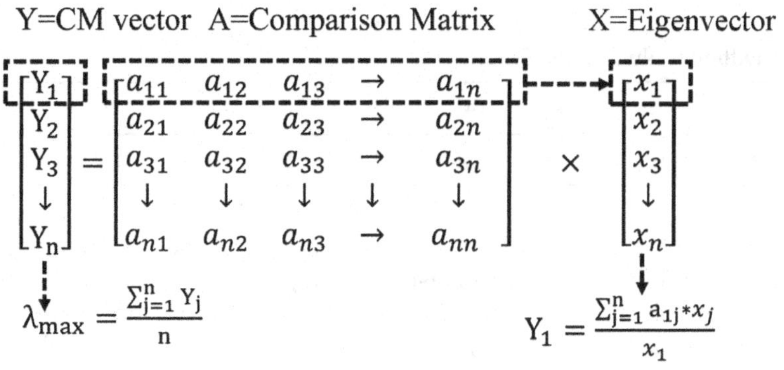

Y=CM vector A=Comparison Matrix   X=Eigenvector

$$\begin{bmatrix} Y_1 \\ Y_2 \\ Y_3 \\ \downarrow \\ Y_n \end{bmatrix} = \begin{bmatrix} a_{11} & a_{12} & a_{13} & \rightarrow & a_{1n} \\ a_{21} & a_{22} & a_{23} & \rightarrow & a_{2n} \\ a_{31} & a_{32} & a_{33} & \rightarrow & a_{3n} \\ \downarrow & \downarrow & \downarrow & \downarrow & \downarrow \\ a_{n1} & a_{n2} & a_{n3} & \rightarrow & a_{nn} \end{bmatrix} \times \begin{bmatrix} x_1 \\ x_2 \\ x_3 \\ \downarrow \\ x_n \end{bmatrix}$$

$$\lambda_{max} = \frac{\sum_{j=1}^{n} Y_j}{n} \qquad\qquad Y_1 = \frac{\sum_{j=1}^{n} a_{1j}*x_j}{x_1}$$

FIGURE 6.3
The measurement of the consistency measure.

then divided by the component in the eigenvector corresponding to M_j. The procedure to find the *CM* is shown in Figure 6.3. The *CM* vector is averaged for computing λ_{max}.

$$Y_j = \frac{M_j * X}{r_i}, \text{where } j = 1,2,3,\ldots,n \tag{6.6}$$

$$\lambda_{max} = \frac{1}{n}\sum_{j=1}^{n} Y_j \tag{6.7}$$

Consistency Index: CI denotes the deviation or the inconsistency [54] of the pairwise comparison matrix for an element. The *CI* of the pairwise comparison matrix for the F_1 criterion is calculated using Equation (6.8) by putting the value of λ_{max}. The value of $\lambda_{max} = 6.07$ and $n = 6$ are put in Equation (6.8).

$$CI = \frac{(\lambda_{max}-n)}{(n-1)} \tag{6.8}$$

In Equation (6.8), n represents the criterion number for controller selection in the comparison matrix. Here, six alternatives are considered; therefore, n is equal to 6. The resultant value for $CI = 0.01$, according to Equation (6.8).

Consistency Ratio: The reliability of the pairwise comparison matrix is verified by calculating the *CR* value. The *CR* is calculated according to Equation (6.9). In Equation (6.9) the ratio index *(RI)* denotes the index ratio. The value of $RI = 1.24$ is derived from Table 6.4, based on the order of the matrix. If the rank of the matrix is three (the actual number of alternatives being compared), then a value corresponding to three is selected for *RI*. In this case, the number of criteria under consideration is 6. Therefore, a value corresponding to 6 will be inserted from Table 6.4. The *CR* is derived by putting *CI* value from Equation (6.8) into Equation (6.9).

$$CR = \frac{CI}{RI} \tag{6.9}$$

TABLE 6.4

Ratio Index for Different Number of Criteria

Criteria	1	2	3	4	5	6	7	8	9	10
Ratio Index	0.00	0.00	0.58	0.90	1.12	1.24	1.32	1.41	1.45	1.49

The CR value is 0.09. A CR of 0.1 or less is accepted for the inconsistent judgments of the comparison matrix; otherwise, the inconsistency is high and pairwise judgments must be made again to satisfy the condition, i.e., $CR \leq 0.1$. The alternatives are pairwise compared to the remaining criteria, i.e., F_2, F_3, F_4, F_5, F_6, F_7, F_8, F_9, and F_{10}. The CI and CR values are computed using the same process for each of these matrices. The CR value is shown in each matrix.

Likewise, the eigenvectors corresponding to X_1, X_2, X_3, X_4, X_5, X_6, X_7, X_8, X_9, and X_{10} are computed corresponding to features F_1, F_2, F_3, F_4, F_5, F_6, F_7, F_8, F_9, and F_{10} respectively along with their CR values. X_1 represents the eigenvector corresponding to the F_1 criterion. Similarly, X_2 represents the eigenvector for the F_2 criterion, X_3 for F_3 and so on. The CR value for calculating each eigenvector is verified to be less than 0.1.

6.4.3 Pairwise Comparison for Criteria with Respect to Controllers

The ten features F_1, F_2, F_3, …, F_{10} of the criteria are pairwise compared for all alternatives C_1, C_2, C_3, C_4, C_5, and C_6. The corresponding eigenvectors for these alternatives are calculated, i.e., X_{11}, X_{12}, X_{13}, X_{14}, X_{15}, and X_{16}. The eigenvectors for alternatives should be calculated using similar calculations as we have done for criteria elements. The CM, CI, λ_{max} and the CR values for each matrix were calculated. The CR value for each eigenvector was checked and verified to be less than 0.1 for maintaining consistency among judgments.

6.4.4 Weighted Super-Matrix

The eigenvectors are computed (which indicates the weight of each criterion in relation to each option and vice versa). We then combine them to form an unweighted super-matrix. Additionally, the unweighted super-matrix is modified to be column stochastic, so that the total of each column equals one. This operation converts the matrix to a super-matrix with weights. The weighted super-matrix illustrates the comparison of criterion alternatives and vice versa. The unweighted super-matrix is identical to the weighted super-matrix; however, the weighted super-matrix is column stochastic. X_1, X_2, X_3, X_4, X_5, X_6, X_7, X_8, X_9, and X_{10} are the eigenvectors respectively for features (i.e., F_1, F_2, F_3, F_4, F_5, F_6 F_7, F_8, F_9, F_{10} which denote the priority values of the features) for each controller. Similarly, X_{11}, X_{12}, X_{13}, X_{14}, X_{15}, X_{16}, the eigenvectors corresponding to C_1, C_2, C_3, C_4, C_5, and C_6, show the priority of the alternatives (controllers) regarding each feature. Finally, we compute the limit super-matrix in order to derive the final weights for the alternatives in the ANP model given in the next paragraph.

6.4.5 Super-Matrix (Limit) Computation

The weighted super-matrix is processed by increasing its power until it converges to a stable matrix. The stable matrix is also referred to as the limit super-matrix. The limit matrix indicates the relative importance of the alternatives and criteria, i.e., the final prioritized values. As a consequence, the limit matrix comprises the final weights assigned to each

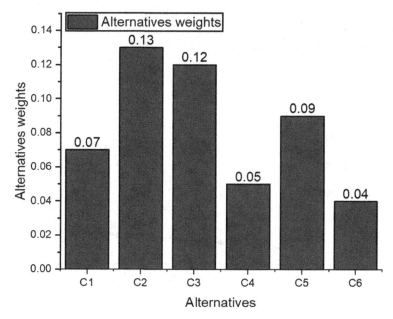

FIGURE 6.4
Priorities of controllers computed by the limit super-matrix.

element in the criterion and alternative clusters. It is computed using a weighted super-matrix in which values are raised to the power of 2^k to produce the identical value for each row, where k is any random integer. The limit super-matrix aggregates all matrices' pairwise comparisons. Additionally, it demonstrates the indirect interaction between the components. The results of the limit super-matrix are shown in Figure 6.4, where heavy weight represents the standing alternative. Figure 6.4 shows the final stable weights of all options. As C_2 has the largest weights, it is the best suited controller. According to their final weights determined from the limit super-matrix, the following controllers are suitable: C_3, C_5, C_1, C_4, and C_6. According to the findings, C_2 has a high weight value, and hence the suggested SDN controller with ANP model corresponds to it. Additionally, the suggested controller's performance is tested in the next part via tests, and then compared to the controller proposed using the AHP model. The following section covers the findings of experimental simulations for both AHP and ANP controllers.

6.5 Results and Discussion

The performance of the C_2 controller (ODL) calculated using the suggested technique, i.e., ANP, is to be analyzed. Additionally, performance is compared to that of the AHP-generated Ryu controller. The Mininet Python API was used to simulate the network topologies calculated using the proposed and AHP approaches. This network simulator has been frequently used to prototype experiments based on SDN. Mininet 2.3.0d1 and OpenvSwitch (OVS) 2.5.4 were installed in Ubuntu 16.04 LTS. Additionally, the Xming server was launched in order to create and visualize traffic between the source and destination servers.

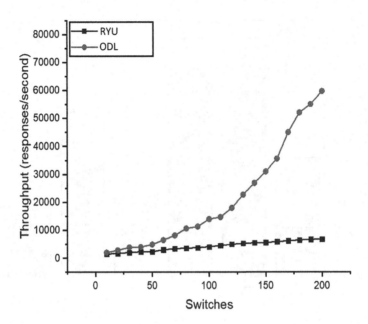

FIGURE 6.5
The throughput evaluation with scalability of the switches in SDN.

6.5.1 Throughput Evaluation

Throughput is determined by sending PACKET-IN packets to the controller and computing PACKET-OUT (responses/second) packets using Cbench. The number of MACs per switch is limited to 2000, the switches are changed between 100 and 200, and each test is repeated ten times. The findings indicate that the ANP controller's throughput does not diminish and that it starts quickly, as seen in Figure 6.5.

6.5.2 Utilization of CPU

To determine the CPU consumption, we utilized a program called sysbench and evaluated both ODL and Ryu controllers. The findings for CPU use at 20-second intervals are shown in Figure 6.6. The graph demonstrates that at peak controller use, utilization does not exceed 45% for a controller with ANP and 30% for a controller with AHP. However, with typical traffic, the controller suggested with AHP achieves a maximum utilization of 19% while the controller proposed with ANP achieves a maximum utilization of 26%.

6.6 Conclusion and Future Scope

The primary objective of this study was to examine numerous strategies for selecting controllers for SDN. The selection of a controller is influenced by a number of factors, including platform support, northbound and southbound interfaces, productivity, and modularity. As a result, we defined it as an MCDM issue. Additionally, the ANP MCDM was used to resolve this issue. To begin, we determined the characteristics that affect performance,

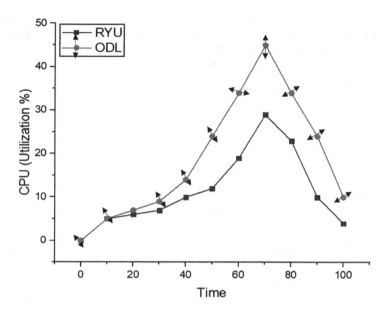

FIGURE 6.6
Percentage of CPU utilization.

namely the criterion parameters and the alternatives (controllers). Then, pairwise comparisons between each characteristic in the criteria cluster and each alternative in the controller's cluster were performed, and vice versa. We obtain priority vectors (eigenvectors) from these comparisons, which are then put in an unweighted super-matrix that has been column stochastically transformed to produce a weighted super-matrix. The final matrix, which indicates how the options and criteria are prioritized, is a limit super-matrix. The limit super-matrix findings indicate that the C_2 controller (ODL) has the best feature set among the SDN controllers investigated in this research. Thus, a high-priority or high-weight controller from the limit super-matrix was chosen for further experimental study.

To validate the performance of the feature-based optimal controller, i.e., ODL, we compared it to an AHP-based controller for the same feature set. We compared the two controllers experimentally by examining several QoS indicators, such as CPU use and throughput. We confirmed the experimental results using Mininet, demonstrating that ODL (ANP) beats Ryu (AHP). We generated a controller with a higher priority weight for supporting features than previous controllers using the suggested technique, and the experimental study confirmed in Mininet demonstrated an increase in performance with the ANP controller.

In future, we tend to investigate other MCDM methods for performance benchmarking of SDN controllers. Hence, a comparison will be made with other MCDM schemes to rank the controllers.

References

[1] J. Ali, and B. H. Roh, "An effective hierarchical control plane for software-defined networks leveraging TOPSIS for end-to-end QoS class-mapping," *IEEE Access*, vol. 11, no. 8, pp. 88990–9006, 2020 May.

[2] K. S. Sahoo, M. Tiwary, B. Sahoo, B. K. Mishra, S. RamaSubbaReddy, and A. K. Luhach, "RTSM: response time optimisation during switch migration in software-defined wide area network," *IET Wireless Sensor Systems*, vol. 10, no. 3, pp. 105–111, 2020.

[3] K. S. Sahoo, D. Puthal, M. S. Obaidat, A. Sarkar, S. K. Mishra, and B. Sahoo, "On the placement of controllers in software-defined-WAN using meta-heuristic approach." *Journal of Systems and Software*, vol. 145, pp. 180–194, 2018.

[4] S. Nithya, et al., "SDCF: A software-defined cyber foraging framework for cloudlet environment," *IEEE Transactions on Network and Service Management*, vol. 17, no. 4 pp. 2423–2435, 2020.

[5] B. A. A. Nunes, M. Mendonca, X. N. Nguyen, K. Obraczka, and T. Turletti, "A survey of software-defined networking: Past, present, and future of programmable networks," *IEEE Communications Surveys & Tutorials*, vol. 16, no. 3, pp. 1617–1634, 2014.

[6] H. Farhady, H. Y. Lee, and A. Nakao, "Software-defined networking: a survey," *Computer Networks*, vol. 81, no. C, pp. 79–95, 2015.

[7] N. Mc J. H. Cox et al., "Advancing software-defined networks: A survey," *IEEE Access*, vol. 5, pp. 25487–25526, 2017.

[8] N. McKeown, T. Anderson, H. Balakrishnan, G. Parulkar, L. Peterson, J. Rexford, S. Shenker, and J. Turner, "Openflow: enabling innovation in campus networks," *ACM SIGCOMM Computer Communication Review*, vol. 38, no. 2, pp. 69–74, 2008.

[9] J. Ali, G. M. Lee, B. H. Roh, D. K. Ryu, and G. Park, "Software-defined networking approaches for link failure recovery: A survey," *Sustainability*, vol. 12, no. 10, p. 4255, 2020.

[10] J. Ali, B. Roh, and S. Lee, "QoS improvement with an optimum controller selection for software-defined networks," *PLoS One*, vol. 14, no. 5, p. e0217631, 2019.

[11] J. Ali, and B. H. Roh, "Quality of service improvement with optimal software-defined networking controller and control plane clustering," *Computers, Materials and Continua*, vol. 67, no. 1, pp. 849–75, 2021 Jan 1.

[12] J. Ali, S. Lee, and B. H. Roh, "Performance analysis of POX and Ryu with different SDN topologies," in *Proceedings of the 2018 International Conference on Information Science and System* (pp. 244–249), 2018 Apr 27.

[13] S. J. Vaughannichols, "OpenFlow: The next generation of the network," *Computer*, vol. 44, no. 8, pp. 13–15, 2011.

[14] J. Crawshaw, "NETCONF/YANG: What's holding back adoption & how to accelarate it," *Heavy Reading Reports*, 2017.

[15] A. Lara, A. Kolasani, and B. Ramamurthy, "Network innovation using OpenFlow: A survey," *IEEE Communications Surveys and Tutorials*, vol. 16, no. 1, pp. 493–512, 2014.

[16] Open Networking Foundation. Available online: https://www.opennetworking.org, 2013.

[17] P. Manso, J. Moura, and C. Serro, "SDN-based intrusion detection system for early detection and mitigation of DDoS attacks," *Information*, vol. 10, pp. 106, 2019.

[18] N. Gude et al., "NOX: Towards an operating system for networks," *ACM SIGCOMM Comput. Commun. Rev.*, vol. 38, no. 3, pp. 105–110, 2008.

[19] POX Network Software Platform. n.d. https://github.com/noxrepo/nox

[20] Ryu SDN Framework. Ryu. https://osrg.github.io/ryu/

[21] Floodlight SDN OpenFlow Controller. n.d. https://github.com/floodlight/floodlight

[22] Trema. n.d. https://github.com/trema/trema

[23] OpenDaylight Foundation. n.d. OpenDayLight. http://www.opendaylight.org/

[24] Open Network Operating System. n.d. https://github.com/opennetworkinglab/onos

[25] G. Wang, Y. Zhao, J. Huang, and Y. Wu, "An Effective Approach to Controller Placement in Software Defined Wide Area Networks," *IEEE Transactions on Network and Service Management*, vol. 15, no. 1, 2018.

[26] P. Bispo, D. Corujo, and R. L. Aguiar, "A qualitative and quantitative assessment of SDN controllers," in *Young Engineers Forum (YEF-ECE) 2017 International*, pp. 6–11, 2017.

[27] R. Izard, and H. Akcay, "Floodligh web GUI," https://floodlight.atlassian.net/wiki/spaces/floodlightcontroller/pages/40403023/Web+GUI, 2017.

[28] R. Khondoker, A. Zaalouk, R. Marx, and K. Bayarou, "Feature-based comparison and selection of software defined networking (SDN) controllers," in *2014 World Congress on Computer Applications and Information Systems (WCCAIS)*, pp. 1–7, 2014.

[29] O. Belkadi, and Y. Laaziz, "A systematic and generic method for choosing A SDN controller," *International Journal of Computer Networks and Communications Security*, vol. 5, no. 11, pp. 239–247, 2017.

[30] A. Abdelaziz, A. T. Fong, A. Gani, U. Garba, S. Khan, A. Akhunzada, H. Talebian, and K. W. R. Choo, "Distributed controller clustering in software defined networks," *PLoS One*, 2017.

[31] J. Huang, J. Zou, and C. C. Xing, "Competitions among service providers in cloud computing: A new economic model," *IEEE Transactions on Network and Service Management*, vol. 15, no. 2, pp. 866–877, 2018.

[32] J. Huang, Q. Duan, S. Guo, Y. Yan, and S. Yu, "Converged network-cloud service composition with end-to-end performance guarantee," *IEEE Transactions on Cloud Computing*, vol. 6, no. 2, pp. 545–557, 2018.

[33] D. Anderson, "An investigation into the use of software defined networking controllers in aerial networks," in *IEEE Military Communications Conference (MILCOM)*, 2017.

[34] O. Salman, I. H. Elhajj, A. Kayssi, and A. Chehab, "SDN controllers: A comparative study," in *2016 18th Mediterranean Electrotechnical Conference (MELECON)*, pp. 1–6, 2016.

[35] Y. Zhao, L. Iannone, and M. Riguidel, "On the performance of SDN controllers: A reality check," in *IEEE Conference on Network Function Virtualization and Software Defined Network (NFV-SDN)*, 2015.

[36] A. Shalimov, D. Zuikov, D. Zimarina, V. Pashkov, and R. Smeliansky, "Advanced study of SDN/OpenFlow controllers," in *Proceeding 9th Central Eastern European Software Engineering Conference Russia (CEE-SECR)*, 2013.

[37] A. L. Stancu, S. Halunga, A. Vulpe, G. Suciu, O. Fratu, and E. C. Popovici, "A comparison between several Software Defined Networking controllers," in *12th International Conference on Telecommunication in Modern Satellite, Cable and Broadcasting Services (SIKS)*, pp. 223–226, 2015.

[38] K. Kaur, S. Kaur, and V. Gupta, "Performance analysis of python based OpenFlow controllers," in *EEECOS*, 2016.

[39] I. Z. Bholebawa, and U. D. Dalal, "Performance analysis of SDN/OpenFlow controllers: POX versus floodlight," *Wireless Personal Communications*, vol. 28, no. 2, pp. 1679–1699, 2018.

[40] H. Shiva, and C. G. Philip, "A comparative study on software defined networking controller features," *International Journal of Innovative Research in Computer and Communication Engineering*, vol. 4, no. 4, 2016.

[41] V. R. S. Raju, "SDN controllers comparison," in *Proceedings of Science Globe International Conference*, 2018.

[42] D. Sakellaropoulou, "A qualitative study of SDN Controllers," https://mm.aueb.gr/master_theses/xylomenos/Sakellaropoulou_2017.pdf, 2017.

[43] A. A. Semenovykh, and O. R. Laponina, "Comparative analysis of SDN controllers," *International Journal of Open Information Technologies*, vol. 6, no. 7, 2018.

[44] G. A. Miller, "The magical number seven, plus or minus two some limits on our capacity for processing information," *Psychological Review*, vol. 63, no. 2, pp. 81–97, 1956.

[45] D. Erickson, "The beacon OpenFlow controller," in *Proceedings of ACM SIGCOMM Workshop on Hot Topics in Software-defined Networking (HotSDN)*, 2013.

[46] Z. Cai, A. L. Cox, and T. S. Ng, "Maestro: A system for scalable OpenFlow control," Tech. Rep. TR10-08, Rice University, 2010.

[47] http://opennetworkingfoundation.github.io/libfluid/

[48] L. Mamushiane, A. Lysko, and S. Dlamini, "A comparative evaluation of the performance of popular SDN controllers," in *10th Wireless Days Conference (WD)*, 2018.

[49] A. Ishizaka, and P. Nemery, *Multi-criteria decision analysis: methods and software*, John Wiley & Sons, 2013.

[50] G. Buyukozkan, C. Kabraman, and D. Ruan, "A fuzzy multi-criteria decision approach for software development strategy selection," *International Journal of General Systems*, vol. 33, pp. 259–280, 2004.

[51] G. A. Mendoza, and H. Martins, "Multi-criteria decision analysis in natural resource management: A critical review of methods and new modelling paradigms," *Forest Ecology and Management*, pp. 1–22, 2006.

[52] X. Yan, P. Dong, T. Zheng, and H. Zhang, "Fuzzy and Utility Based Network Selection for Heterogeneous Networks in High-Speed Railway," in *Wireless Communications and Mobile Computing*, 2017.

[53] P. Boateng, Z. Chen, and S. O. Ogunlana, "An analytical network process model for risks prioritisation in megaprojects," *International Journal of Project Management*, vol. 33, no. 8, pp. 795–811, 2015.

[54] S. Nazir, S. Anwar, S. A. Khan, S. Shahzad, M. Ali, R. Amin et al., "Software component selection based on quality criteria using the analytic network process," *Abstract and Applied Analysis*, 2014.

[55] H. Farman, H. Javed, B. Jan, J. Ahmad, S. Ali, F. N. Khalil, and M. Khan, "Analytical network process based optimum cluster head selection in wireless sensor network," *PLoS One*, vol. 12, no. 7, 2017.

[56] S. Sun, Y. Wu, Y. Ma, L. Wang, Z. Gao, and C. Xia, "Impact of degree heterogeneity on attack vulnerability of interdependent networks," *Scientific Reports*, 2016.

[57] T. L. Saaty, *Decision making with dependence and feedback: The analytic network process*, RWS Publications, 2001.

7

A Scalable Software-Defined Edge Computing Model for Sustainable Smart City Internet of Things (IoT) Application

Hemant Kumar Apat, Bibhudatta Sahoo and Sagarika Mohanty
National Institute of Technology, Rourkela, India

Kshira Sagar Sahoo
SRM University, Andhra Pradesh, India
Umea° University, Umea° 901 87, Sweden

CONTENTS

DOI: 10.1201/9781003213871-7

7.1 Background and Motivation

Over the past couple of years, the demand for computation resources has been increasing due to the explosive growth of internet of things (IoT) devices and the emergence of a variety of IoT applications in such domains as smart healthcare, smart city, intelligent transport, smart games, and many more. The limited bandwidth and high latency existing in the traditional cloud computing system cannot meet user requirements, although the integration of IoT and cloud offers elastic resource management with robust pay-per-use policies. The data-intensive, latency-sensitive, latency-tolerant characteristic of multiple IoT applications pose major challenges to the current state-of-the-art centralized cloud architecture in terms of quality of service (QoS). To meet the computational requirements of various geo-distributed heterogeneous real-time IoT applications we propose an edge computing paradigm referred to as fog computing to address the existing challenges in an ad-hoc manner, such that users experience better QoS. The fog computing model is a virtualized decentralized model consisting of heterogeneous fog devices to execute the services brought from the cloud layer. The fog layer includes a variety of hardware devices including smartphones, desktops, base stations, or any smart switches. Resources in the fog layer are highly distributed, and hence we here integrate the advanced concept of the software-defined network in the fog layer whose primary function is to allocate various IoT tasks so that fewer IoT applications are offloaded to the cloud computing layer [1, 2]. An SDN controller is placed between fog and cloud layers for the efficient routing of the IoT data to avoid overloading or underloading fog devices [3–5]. In this article, we address the challenges of the existing cloud computing model and propose an SDN-enabled fog computing model that supports a variety of IoT applications in a smart city. Task allocation in multiple clouds is considered one of the best choices for improving various QoS issues [5, 6]. The edge devices allow different computing facilities to deliver the services to the IoT users in minimum latency. However, there are many challenges in edge computing scenarios, including how to allocate the resources to the geo-distributed IoT applications in such a way as to achieve specific QoS. Other challenges arise due to heterogeneous IoT devices and the unscheduled and dynamic requests they mostly generate. Various advanced networking technologies like network function virtualization (NFV) and software-defined networking (SDN) need to be optimized in order to analyze traffic from sensor data to take intelligent decisions. OpenFlow protocol allows SDN to decouple the control plane and forwarding plane into a separate entity for taking centralized control action for incoming IoT packets. In SDN the programmed network controller consists of the various application that runs on top of a virtual machine to support various services for IoT applications in an IoT network.

7.2 Introduction

The internet of things (IoT) is rapidly evolving as the novel paradigm that interconnects a large number of heterogeneous IoT devices to provide different services. The basic idea behind IoT is to extend everything with a computing power and connection to the internet and enable them to sense, compute, communicate and control the surrounding environment. The term IoT was first coined by Kelvin Ashton in 1999 while working with a global network of physical objects connected through radio frequency identification (RFID) in supply-chain management [7]. The development of different real-time applications such as autonomous vehicles, smart healthcare services, etc. requires a high quality of service from network service providers [8]. IoT-enabled devices take data from the environment and forward it to the network layer where multiple devices such as gateway, router, and smart switches are available. These devices have very limited processing capability, hence a network and computing infrastructure is required that is capable of processing these environmental data and giving the result to the end user with minimum latency and fast response time. Initially, cloud computing was the only option available for offloading real-time tasks generated by different sensor devices to facilitate computation and storage resources for underlying devices. However, intelligent IoT applications have been developed consisting of tasks coded with machine learning and deep learning generated through IoT devices. Task offloading requires a quick decision if a particular node has not had enough resources. Today's popular cloud computing model has reached a peak due to its various superior qualities such as salable, ubiquitous services in the form of IaaS, PaaS, and SaaS. The centralized cloud data center provides different services to IoT applications, processing, executing, and sending the result back to IoT user at a minimal cost. However, for some intelligent IoT applications, the cloud computing paradigm is unable to provide the required QoS. A variety of solutions have been proposed such as mist computing, dew computing, mobile cloud computing, edge computing, fog computing, multi-access edge computing (MEC) [9]. Different advanced cloud model like the vehicular cloud have also been proposed by authors to deal with various types of vehicle-to-vehicle (V2V) communications [10–13].

According to IDC forecasts, there will be 41.6 billion connected IoT devices, or "things," generating 79.4 zettabytes (ZB) of data in 2025 [14, 15]. The extraordinary growth of IoT devices and the large volume of data put of a lot of strain on the existing centralized cloud computing network, with undesirable latency due to the long distance between the data generator and cloud data center. Recently the edge computing model has been developed as a potential computing paradigm capable of processing and executing IoT data at the edge of the network with low communication delays and high bandwidth for geo-distributed IoT devices. The reduced latency achieved by moving the resources to the edge enables the response time for latency-sensitive IoT applications to match the application's deadline.

Fog computing is a distributed computing model for processing IoT data close to the source i.e., at the network edge, to improve the latency, bandwidth, and quality of service of various cloud-based IoT applications [16–18]. Candidates for fog devices are various smart devices like SDN switches, base stations, routers, smartphones, and smart vehicles with storage capacity and computing power. In fog computing, the resources are configured in an ad-hoc manner and an application or a collection of applications may privately make use of them. These resources are not publicly available like cloud resources. Nor are they evenly distributed, but are sporadic in their geographic distribution. The first formal definition of fog computing stated that "Fog computing is a highly virtualized platform that provides compute, storage and networking services between end devices and traditional Cloud Computing Data Centers, typically, but not exclusively located at the edge of the network" [16]. As the definition suggests, fog computing does not necessarily replace cloud computing but complements it. Fog computing fills the current gap between cloud and things, by distributing computing and control, storage, and networking and communications functions closer to end-user devices. Essentially, it eliminates/vastly reduces networking loads and latency.

Software-defined networking (SDN) has recently gained much popularity among the industry and research community [19–21]. The roots of SDN are in programmable networks and the separation of control and data planes that provides the network service provider with flexibility and a dynamically customized network. The success and rapid adoption of SDN in today's network are due to the OpenFlow protocol. OpenFlow protocols enable multiple packet handling rules that do not require major hardware upgrades. In SDN-based network, the controller has a global view of network devices. The SDN concept abstracts available network resources and controls them using a centralized and intelligent authority that aims to optimize traffic flow in a flexible manner, which in turn increases availability [17, 22, 23].

Objectives of Chapter

In this chapter, we propose an SDN-based distributed fog computing model with fog control manager as a fog orchestrator node that performs various functions like profiling QoS parameters, offloading decisions, resource management, etc. In addition, the objectives of the chapter are:

(i) A three-layer architecture of the IoT ecosystem is presented with a detailed description of three important IoT entities: IoT devices, IoT applications, and IoT services.
(ii) A mathematical model of different possible smart city IoT applications is elaborated.
(iii) An SDN controller as the middle layer between the fog node and cloud to enabling easy collaboration between the services and IoT applications is introduced.

Organization of Chapter

The chapter is organized as: Section 7.3 highlights literature review. Section 7.4 discusses the evolution of Edge Computing. Section 7.5, 7.6 enlightens Internet of things-general overview and characteristics. Section 7.7. highlights proposed Edge-SDN architecture. Section 7.8 overlays mathematical models. Section 7.9 discusses challenges. Section 7.10 concludes the chapter with future scope.

7.3 Literature Survey

Most work on the development of fog computing for different real-time IoT applications with the integration of SDN [24, 25] only considers applications where the task model is either dependent or independent. The survey in this section covers IoT-cloud and IoT-fog-cloud. We have surveyed various works related to some traditional IoT applications, but there are more IoT applications that have stringent QoS requirements. Sahoo et al. [26] proposed a virtual machine allocation policy in the cloud for a set of heterogeneous tasks defined by size, arrival, and deadline. Works include dealing with system architecture and with different performance measures in such problems as application placement [17], task scheduling [6], resource provisioning [27, 28], and resource allocation methods [29] [30–32].

Ghobaei-Arani et al. [29] classified resource management issues in fog computing into application placement, resource scheduling, task offloading, load balancing, and resource allocation. However, the survey does not cover resource discovery, without which resources cannot be optimally allocated.

Sharma et al. [33] proposed a sustainable edge computing model, SoftEdgeNet, using blockchain technology and SDN to mitigate attacks in the network. SDN is used to provide different real-time analytics using flow rule partition. In addition, an allocation algorithm for multiple IoT applications at the edge of the network is proposed. A three-layer fog computing model is designed with fixed parameters for heterogeneity of IoT applications, real-time analytics, reduced latency and bandwidth, high energy efficiency, scalability and security.

Jararweh et al. [34] proposed a reliable edge computing framework to address the performance of various smart-city IoT applications, specifically service availability, reliability, sustainability and security. An intrusion detection system is implemented to enhance parameters like reliability and security. Performance of the proposed approaches is evaluated experimentally in densely populated and lightly populated smart cities. The result of the simulation outperforms other state-of-the-art frameworks. It was concluded that, with the growth in the number of IoT applications, mechanisms are needed to provide resources under overload conditions and scalability is one of the important parameters to be addressed for sustainable smart-city applications.

Pekar et al. [35] addressed various IoT application domains for identifying traffic patterns to establish an efficient method to improve IoT network performance. Different IoT application domains are compared on the basis of traffic size, area coverage and their respective quality of service (QoS) demands.

Gu et al. [36] presented cost-efficient resource management in a fog computing framework to support periodic healthcare IoT applications generated by healthcare devices. Healthcare IoT applications are latency sensitive and have specific QoS requirements. The events are generated randomly and chosen as random processes to validate the proposed model. Different constraints are added in the optimization problem to find the minimal cost of executing healthcare IoT applications.

Minh et al. [37] proposed a fog computing framework, Fog Fly, to collect real-time data about traffic conditions, optimizing various parameters like latency, energy and operational costs for an intelligent transport system. The proposed methods are evaluated with diverse cloud-based computing frameworks. The result reveals the proposed methods outperform existing cloud-based approaches.

Pallewatta et al. [38] proposed a decentralization placement strategy of various interrelated services, called microservices, represented with a directed acyclic graph (DAG) where the vertices represent the microservices and the edges represent dependencies between microservices. Three-layer heterogeneous fog architecture is proposed to process latency-sensitive microservices. Three parameters – latency, critical path and network resource – are used to evaluate the results, which are compared with cloud ward only and edge ward.

From the literature survey, it can be concluded that most of the authors considered IoT applications as latency sensitive or computation intensive, without any mathematical models of these IoT applications. Most research papers added very conventional parameters to evaluate QoS, like latency, response time, and energy; however there are other non-quantitative parameters like scalability, security, and elasticity which are major QoS issues in large-scale IoT applications. Most authors proposed a fog-cloud hierarchical edge computing model to evaluate performance in various types of IoT applications. None of the authors specifically defined the occurrence of the IoT applications, whether periodic or aperiodic.

Other important challenges for IoT applications are the requirement for highly scalable and elastic resource management policies for service providers. Resource scaling is about the decision to scale up or scale down according to the frequency of incoming application requests. There is much ambiguity between IoT services, applications and tasks. The following points must be determined in order to evaluate the performance of the proposed model:

1. What are the different strategies for allocating the various IoT applications to the appropriate IoT service?
2. Who is responsible for the commissioning and decommissioning of services?
3. What is the lifespan of an IoT application?
4. What are the attributes of a general IoT application?
5. Under what conditions is an IoT application deadline specific?

Very few research articles reference queueing theory to estimate the number of IoT tasks executed in a particular time, and assumed as the task arrival pattern is Poisson distribution which is not true in the case of periodic IoT application. The other probability distribution may be significantly important in case of IoT applications. Table 7.1 highlights literature review of Multiple IoT Application Scenarios.

TABLE 7.1

Literature Review of Multiple IoT Application Scenarios

References	Parameters	IoT Application	Framework		Task
Jararweh et al., 2020 [34]	Reliability, availability and security	Smart city	Fog-cloud		Delay sensitive
Sharif et al., 2020 [39]	Delay, energy and resource utilization	Smart surveillance system	Fog-cloud		I/O intensive
Xia et al., 2020 [40]	Task completion Ratio and service delay	Artificial Intelligence-based IoT application	Edge	Cloud Computing	Resource intensive with deadline
Minh et al., 2018 [37]	Energy	Intelligent transport system	Fog-cloud		Delay sensitive with deadline
Farahani et al., 2018 [41]	Scalability and security	Smart healthcare	Fog-cloud		Accuracy sensitive
Korala et al., 2020 [42]	Delay	ITS	Edge-cloud		Delay sensitive
Venticque et al., 2019 [43]	Delay, response time and energy	Smart energy	Fog-cloud		Computation intensive
Teng et al., 2019 [44]	Scalability, power consumption, response time	Smart city	Fog-cloud		Resource intensive without deadline
Arora et al., 2020 [45]	Response time, RU, and energy	Smart city	Fog-cloud		Computation intensive no deadline
Ahmed et al., 2017 [46]	Latency	IoT big data	Mobile edge computing		Computation intensive no deadline
Rahman et al., 2019 [47]	Big data analytics	Industrial IoT	Edge computing		IoT big data

7.4 Evolution of Edge Computing

7.4.1 Cloud Computing

Cloud computing is a distributed system that offers on-demand services to end users through the pool of computing resources. It offers various services such as infrastructure as a service (IaaS), platform as a service (PaaS), and software as a service (SaaS) to users at minimal cost [48–50]. However the large volume of data generated by IoT devices increases latency and network congestion in the cloud data center, affecting the quality of services to some mission-critical application like healthcare and smart traffic systems. The large number of computational servers inside the cloud date center also increase energy consumption and CO_2 emissions. Advanced networking concepts have been developed such as software-defined networking (SDN) controllers that manage IoT traffic so that each application obtains a result before its deadline. The other limitation of cloud computing are as follows:

Security and Privacy: Most IoT devices offload the data to the cloud data center for processing and execution due to their low storage and low computational capabilities. As most of the cloud is now public and maintained by different cloud service providers, privacy and security depend on third-party vendors, which is a major loophole in cloud computing systems.

Reliability: As IoT devices partially offload data for different applications, the processing and execution of the data depends on the cloud resources. This is problematic for emergency applications like smart healthcare and smart video surveillance systems.

7.4.2 Cloudlet Computing

Cloudlets are small servers or groups of servers that have sufficient resources and are placed at least one hop away from mobile users. These servers employ virtual machine (VMs) to execute resource-intensive applications offloaded by mobile users. Due to the rapid development of various latency-sensitive applications with stringent end-to-end delay requirements, cloudlets mostly serve these applications. Their primary objective is to provide real-time interaction with mobile users with minimum distance rather than sending data to a cloud data center.

7.4.3 Mobile Edge Computing (MEC)

The concept of mobile edge computing (MEC) was initiated by the European Telecommunication Standards Institute (ETSI) with the objective of providing cloud-like services closer to mobile devices within radio access network (RAN) [51–53]. MEC services can be set up by various platforms such as LTE base station and 3G radio network. The primary objective of adopting MEC is to improve application performance by reducing the delays occurring in the cloud data center and improve the quality of experience by optimizing the network resources. The types of application benefiting from this technology are augmented reality and video processing.

7.4.4 Edge Computing

Edge computing is a computing model that extends cloud services to the devices at the edge of the network, enabling processing and storage to take place on edge devices. The computation of data generated through IoT devices takes place in close proximity to the data sources [54, 55]. There is some debate as to whether edge computing is the same as fog computing. While most authors have assumed edge computing and fog computing are the same, some have identified characteristic differences, for example, edge computing does not support virtualization, whereas fog computing does.

One reason for choosing fog computing is the availability of standard architecture. According to the Open Fog Consortium, fog computing is a three-layer architecture used to support various services for IoT applications. Other benefits are reduced latency and bandwidth for delay-sensitive and throughput-sensitive IoT applications. To reduce network latency and traffic between end devices, edge nodes and devices with computing power execute a vast number of computer functions (e.g., data processing, temporary storage, device management, decision making, and privacy protection).

The edge computing paradigm presents a number of challenges, including resource management. The resource requirements of different IoT applications vary due to different hardware and software configurations, along with different communication protocols. The successful adoption of edge computing requires resources to be managed effectively and efficiently. The edge devices also suffer from resource scarcity, and IoT application requests are mostly unscheduled in nature. Various resource management strategies are suggested in the literature (e.g., energy-aware resource management, latency-aware resource management, etc.).

7.4.5 Fog Computing

Fog computing is an extension of cloud computing that integrates various edge devices to provide services and resources at the edge of the network with numerous benefits, by minimizing communication delay and computation costs. The devices used in the fog layer may be access points, gateways, and routers for constant supervision of IoT data. The fog computing layer is virtualized and offers various resources inherited from the cloud, such as computation resources, communication resources, and storage resources. The following are among the benefits of fog computing for IoT applications:

Latency Requirement: Latency in the traditional cloud computing model is high for most real-time IoT applications. Fog computing helps to execute services near the data generation layer and hence the transmission latency and propagation latency are lower than with cloud computing. Delays from network congestion are minimized in fog computing. However, there are other challenges associated with the fog computing layer. If resources are not available in the requested fog node or the requested resource is more than is available, fog nodes forward the IoT data to a centralized cloud data center for execution using a multi-hop pathway.

Bandwidth constraints: Fog computing offers the ability to carry out data processing tasks closer to the network edge. The amount of raw data sent to the cloud is reduced by adopting the fog layer as the intermediate layer between the cloud and IoT. Some preprocessing and data aggregation tasks can be performed by the fog nodes to reduce data sent to the cloud, which can reduce bandwidth consumption to some extent.

Energy-constrained IoT devices: Most IoT devices are battery operated and deployed in various geographical locations where charging or providing non-renewable sources of energy is the only option. Execution of IoT data by these devices

sometimes results in cell discharge. Fog computing saves energy by partially executing the application by collaborating with the cloud.

Increased availability: Fog computing is an autonomous framework that does not need
to connect with the cloud if its own resources are sufficient for IoT applications.
The resource demand from IoT devices depends on the frequency of data they
generate and the application used. The availability factor is a major issue in distributed computing architecture, as most requests are unscheduled and can occur
at any time. Fog computing provides greater support for availability.

Mobility support: Most IoT devices, such as smartphones, autonomous vehicles, are in
non-static mode, frequently changing their location, while some end devices, like
cameras in the smart city, remain static. Mobility of the end device is supported by
the fog computing layer.

Better security and privacy: Most IoT devices now used by different smart city applications like smart home, and smart healthcare consist of various sensitive items of
information that must not be processed in the public cloud. In the fog paradigm,
users have the ability to control the collected data. Fog devices reduce the need to
transfer private data to the cloud, keeping it to be processed locally. Fog devices
can perform a wide range of security functions, manage and update the security credentials of constrained devices, and monitor the security status of nearby
devices. Data integrity and privacy are further ensured when data have to travel a
shorter network distance to reach the computation node.

7.5 Internet of Things (IoT) Network

Internet of Things (IoT) is a network that describes the connectivity of physical objects –
"things" – with embedded sensors and software for the purpose of connecting and
exchanging data with other sensors or IoT devices over the internet with minimal human
intervention. The IoT devices may be smartphones, wearable devices or sophisticated
industrial devices consisting of multiple sensors capable of collecting and processing information from the environment [56]. A well-defined IoT architecture is still not established.
However, a three-layer high-level architecture is commonly accepted [37] (see Figure 7.1).
The bottom or end layer is the perception or sensing layer responsible for sensing the
environment where the sensor or IoT device is deployed; the second is the network layer,
connected to the perception layer by various communication protocols such as Zigbee,
Bluetooth Low Energy (BLE), RFID etc.; and the third layer is the application layer, providing application programming interface (API) to the IoT users.

7.5.1 Perception Layer

This layer is responsible for sensing the environment where sensors are deployed. Its
primary aim is to identify the physical objects and process the sensor data. The three
important components are sensors, actuators, and machines. The sensor layer constructs a
straight connection with IoT infrastructure and performs various task like data preprocessing, removing data redundancy, and resolving inter-operability issues. Power consumption for task execution can be predicted by the sensors.

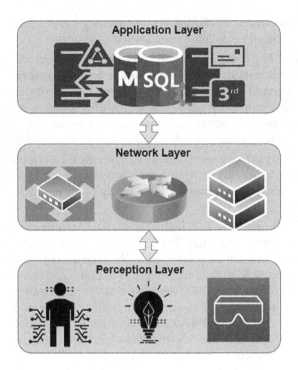

FIGURE 7.1
Three-layer IoT architecture [57].

7.5.2 Network Layer

The network layer of IoT architecture deals with different communication protocols (e.g., IEEE 802.15, Bluetooth Low Energy (BLE), Zigbee, and RFID). IoT network traffic can be classified as (i) throughput and delay-tolerant elastic traffic; and (ii) bandwidth- and delay-sensitive inelastic (real-time) traffic [56]. The QoS parameters in this layer are therefore: (i) communication cost, (ii) traffic cost (iii) communication overheads, (iv) latency time, (v) bandwidth, and (vi) throughput.

7.5.3 Application Layer

The application layer provides the application program interface (API) to the end users for interaction with multiple IoT applications, such as such as smart healthcare and smart transport services, etc. The main objective of this layer is to ensure delivery of IoT services with low latency and high throughput (service reliable). The QoS parameters in this layer are: (i) end-to-end latency, (ii) scalability, (iii) availability, (iv) reliability, (v) throughput, and (vi) cost-effectiveness.

7.6 Characteristics of IoT Network

The following are the characteristics of IoT Network:

Distributivity: The location and generation of IoT data varies, with some devices generating continuous data while others generate data in discrete form.

Inter-operability: Multiple manufacturers are producing sensors and their communication protocols are different. D2D communication is an issue in IoT networks.

Scalability: Billions of physical objects now have inbuilt sensor technology. The IoT system and application interface run on top of the network layer must be capable of managing the large volume of data generated by these devices.

Resource scarcity: The power and computation resources are limited. This puts additional constraints on executing applications over an IoT device.

Security: As most IoT users are connected through a gateway, there is a critical need for lightweight security mechanisms.

7.6.1 Internet of Things (IoT) Applications

The IoT system is a service-oriented architecture that aims to provide different services to the devices connected through it. An IoT ecosystem consists of three entities: IoT devices, IoT services, and IoT applications. The IoT devices are capable of acquiring sensor data which need to be processed and analyzed for different actions. IoT services consist of various types of services or the resources required to execute the IoT applications. IoT applications are software used by the IoT users to achieve specific goals, for example in healthcare IoT applications the sensors attached to a person continuously sense the data and send these data through the device. A generic life cycle of IoT devices, IoT service and IoT application is presented in Figure 7.2.

IoT applications are default software-as-a-service (SaaS) applications that take IoT or sensor data as input and analyze the data using dashboards to for multiple computations for various business activities. Among today's IoT applications are smart home, smart health care, intelligent transport systems, augmented and virtual reality, smart cities, and smart grid, etc. As each application has a different target, they are used in different ways and therefore experience different types of traffic behavior.

7.6.2 Challenges of IoT Applications

The following are the challenges of IoT Applications:

IoT application offloading: A way of executing the IoT application to other available resources in case of non-availability of resources. As fog devices are also resource constrained, it is not advisable to execute all tasks in a single fog device; hence a decision is made where to execute some tasks so that this objective remain satisfied. The application offloading problem is a multi-criterion decision problem in which the controller node take decisions for partial execution. The resource may be computation, communication or storage. When an IoT devices sends a request to the edge layer for processing and execution, the edge layer executes some parts while other parts are executed in the cloud layer.

IoT service deployment: Service deployment means deploying the various services to the computing layer so than when an IoT application sends a request it is able to execute and send the result through the actuator in the minimum time. The services may be computation, infrastructure or platform, depending on the application. Where multiple IoT applications demand an atomic service, deciding which one is to execute is challenging if all the tasks are high priority.

Energy consumption: Most IoT devices are connected wirelessly with inbuilt low-energy batteries. The life of an IoT device is determined by its energy consumption. To minimize energy consumption of IoT devices, computation-intensive tasks are offloaded to the edge layer or cloud layer. However, some preprocessing tasks can be performed by IoT devices. At the edge layer, various heterogeneous edge devices are available with varying power

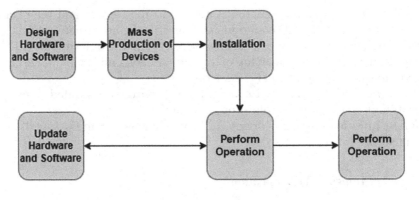

Life Cycle of IoT Device

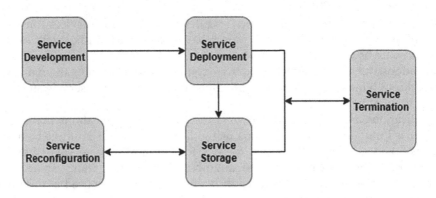

Life Cycle of IoT Service

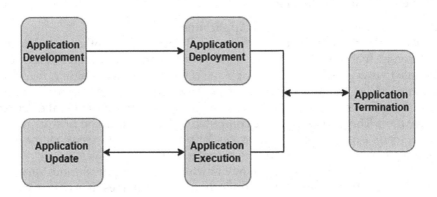

Life Cycle of IoT Application

FIGURE 7.2
A generic life cycle of IoT devices, IoT service and IoT application.

consumption rates. To minimize the energy consumption of the edge layer, it is important to quantify the power consumption of each edge device in both active and idle states.

Availability: Availability is one of the major challenges in IoT as the requests generated by IoT devices are mostly unscheduled and resource demand is highly heterogeneous. Different resources are required – computation, communication and storage – and the computing framework should have sufficient resources to obtain quick results.

Inter-operability: Challenging inter-operability issues arise from different non-adaptive communication protocols, and protocol conversion software must be deployed to resolve these.

Quality of service (QoS): QoS is a major issue, especially in the context of IoT applications. Most studies list conventional QoS parameters for IoT applications. However, there must be a process for categorizing QoS parameters – such as time sensitive, I/O intensive, resource intensive – for various IoT application scenarios (different traffic patterns, varied communication protocols).

Privacy and security: IoT is constantly growing due to the emergence of various forms of wireless technology, such as RFID, WSN and cloud services. IoT devices connected through the gateway continuously send sensor data to the gateway for analysis. The sensor data may include some private health or financial data which requires encryption. Different types of potential threats in the IoT relate to hardware, network, and communication.

7.6.3 Classification of IoT Applications

The different types of IoT applications in use today, are illustrated Figures 7.3–7.6 respectively. However, due to the emergence of machine learning and artificial intelligence their behavior and purpose is not predicted accurately. For example, in a smart city different high-definition IP cameras are now used to monitor certain areas, specifically to observe events such as traffic accidents, thefts from ATM machines, or terrorist activity. The variation of these IoT applications creates many challenges for the IoT service provider.

In order to predict the behavior of various IoT applications, a classification is necessary for most service providers.

Monolithic Applications: An application generated by IoT devices with single tasks that can be executed over a single virtual machine are referred to as monolithic IoT applications. In this category all the IoT application modules are tightly coupled i.e., individual tasks cannot be executed in parallel. A monolithic IoT application is depicted in Figure 7.3.

Independent Applications: In this case IoT applications consist of a finite number of indepennt tasks, and the execution of each task is independent of the execution of other tasks, so they can be executed in parallel. Each task has some specific function for an application, for example, in a body sensor network different sensors are attached to the patient's body to acquire environmental as well as physical data from the patient. The execution tasks for these data are independent, hence they can run in different virtual machines. An independent IoT application structure is shown in Figure 7.4.

Dependent applications: In this category of IoT applications, there are some dependencies between different tasks of an application, i.e., the execution of one task depends on the execution of others. Generally, these types of IoT application are represented by a directed acyclic graph (DAG), where the vertices represent the task of an application and the edges represent the communication delay between the two tasks. A DAG-based IoT application is referred as an IoT workflow. Each IoT application may have one root node or many root nodes; similarly, there are one or more exit nodes. In order to execute the DAG workflow, a priority needs to be assigned to each task. Upward rank and downward rank are calculated for each and every node. There are various methods to sort the DAG nodes where

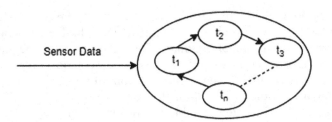

FIGURE 7.3
Monolithic IoT applications.

FIGURE 7.4
Independent IoT applications.

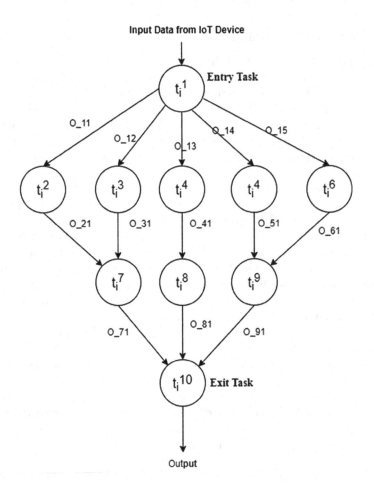

FIGURE 7.5
A simple IoT workflow.

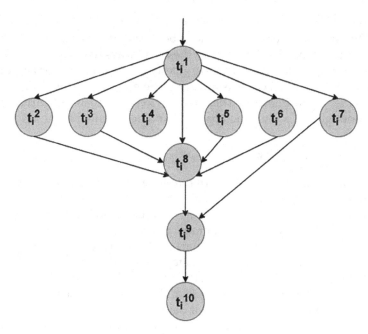

FIGURE 7.6
Different possible DAG of an IoT workflow.

there is no cycle, including topological sorting, breadth first search, and depth first search. For example, in a smart health monitoring application, a patient is monitored remotely through IoT applications, and this can be achieved by executing many tasks like calculating blood sugar, body temperature and environment temperature, etc. These tasks have some dependency, so in an emergency the order of execution is important in order to obtain proper medical advice. A simple dependent task model of an IoT application is depicted in Figure 7.5. An alternative possible DAG is shown in Figure 7.6.

7.7 Proposed Edge-SDN Architecture

A generic three-tier fog-cloud computing architecture inherited from the Open Fog Consortium can process sensor data generated through a variety of IoT devices to support large-scale IoT applications. Fog devices inside the layer – which are connected to the IoT devices and cloud data center through the gateway – are highly virtualized and capable of executing latency-sensitive IoT data. The cloud layer is responsible for processing latency-tolerant and storage-sensitive IoT applications for analytical purposes. Fog devices may be any physical device that has some computing capability to compute IoT tasks. When an IoT device sends a request to the fog device, the device computes the task, sends the result to the end user using the actuators and at the same time stores the data in the cloud server for future use. The fog node inside the fog computing layer can independently execute IoT applications locally, or it may partially execute some tasks and forward other, resource-intensive tasks to the cloud for further execution. Applications requiring low latency are executed in the fog node. Tier 1 consists of various geo-distributed IoT devices that produce a large amount of IoT data which is sent to the IoT gateway – called the fog gateway in our architecture. This gateway works as the bridge between devices, sensors and various items of equipment, systems and cloud. However, these gateways have other functions, like local processing and storage of IoT data. The fog gateway is responsible for aggregation, filtering and analysis of data. Tier 2 consists of various heterogeneous fog devices, according to the service provider's requirements. These fog devices are capable of partial execution and are intelligent enough to decide which parts should be executed in the cloud.

7.7.1 Different Components of Fog-SDN Architecture

The different components of Fog-SDN Architecture, represented in Figure 7.7, are illustrated as follows:

IoT device layer: The bottom layer of a fog computing architecture consists of various geo-distributed IoT devices and is generally referred to as the data generation layer. Devices include mobile phones, laptops, smart grids, smart cards, smart home appliances, and so on. Many potential devices can act as fog devices, like surveillance cameras, Cisco Unified Computing System (UCS) servers, base station (BS) or autonomous vehicles. They are responsible for the sensing, transmission and processing of sensed data to the higher layer, using physical entities or usable event data.

Fog computing layer: The next layer up is the fog computing layer, connected through the gateway to the various geo-distributed IoT devices at least one hop away. This layer is responsible for processing and executing the IoT data forwarded by the respective IoT devices. Fog devices include routers, switches, gateways and other networking elements for processing and storage of IoT data for various latency-sensitive applications. Fog devices support various virtualizations to support multi-tasking and multi-programming, and other network devices which incorporate some intelligence. Terminal devices can easily connect to any fog devices in the nearest network. A fog node is a virtualized device capable of running multiple virtual machines.

Cloud computing layer: The cloud computing layer is the abstraction layer consisting of single or multiple data centers with one or many cloud servers to execute the various resource-intensive IoT applications. The fog layer is connected to the cloud server using the software-defined network (SDN) switches. The cloud server has many powerful servers and racks with various SDN-enabled devices for analyzing packet traffic in order to allocate resources efficiently.

Fog control manager (FCM): FCM is a controller node that controls the particular cluster in the fog layer. FCM acts like a fog "orchestrator node", capable of executing the IoT task by itself or forwarding the incoming tasks to fog devices deployed in the respective cluster. A queue is maintained by the FCM to analyze incoming requests from IoT devices. FCM consists of various modules like scheduler, resource manager, decomposer, and classifier. In the proposed architecture, one FCM for each cluster is considered. Every FCM in the cluster can easily communicate with other cluster FCMs with a negligible delay. The delay between fog device and FCM is also negligible. As IoT requests are randomly generated, the arrival pattern of incoming tasks takes the form of a Poisson distribution. All the information regarding the fog device, such as computational capacity, available resource, overloaded status, underloaded status, and so on, is stored in the FCM. When an IoT application request is initiated from various IoT devices, the fog devices forward the request to the FCM. A global queue is maintained by the FCM to analyze the incoming request. If the

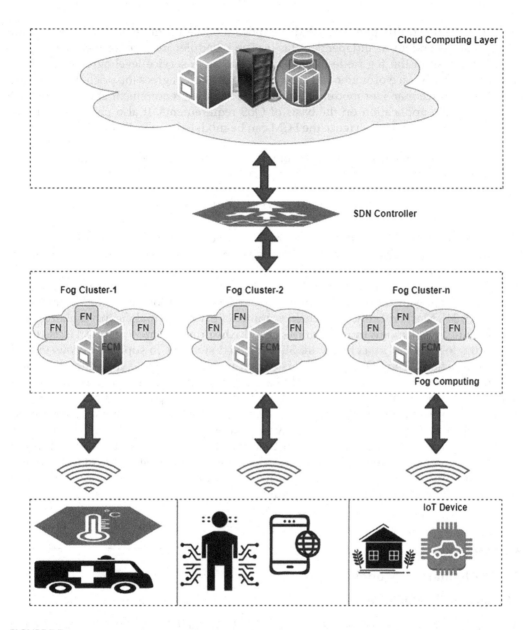

FIGURE 7.7
SDN-edge computing architecture.

queue is full the incoming request tasks are weighed up and an appropriate strategy for forwarding the tasks to the available fog device is chosen. If the queue is full and all the fog devices are overloaded, the FCM of the current cluster communicates with a neighboring FCM to keep some free resource to execute the IoT tasks. Specifically, a fog orchestrator node can provide an inter-operable resource pool, deploy and schedule resources to the application workflow and control QoS. A classifier classifies tasks according to their requirements and traffic pattern. The decompose module partitions the IoT application into sets of dependent or independent tasks. The scheduler module decides whether to process the task in the fog node or cloud data center. A service level agreement (SLA) between the fog service provider and cloud service provider agrees the pricing policy and QoS. The resource manager module estimates the resource requirement and allocation of resources to the application on the basis of QoS requirements. It also performs various resource optimization tasks. Hence the FCM can be modeled as a birth–death process:

birth rate (arrival rate i.e., the number of tasks currently being executed or consuming resources after tasks have been deleted or sent to the cloud server) = λ;

$$\text{death rate (service rate)} = \mu$$

SDN controller: A software-defined controller is used to separate the control plane from the data plane. The SDN controller maintains the networking resources in a centralized way. The controller maintains a general view of the underlying network resources through its logically centralized structure [29]. This simplifies network management, improves resource capabilities, and reduces complexity barriers by employing resources more efficiently [33, 34]. Most crucially, SDN enables immediate decision making in a dynamic context by continuously monitoring network status. In the proposed architecture, it is assumed that all the fog devices in a cluster are SDN-enabled switches to support various virtual network functions. The SDN controller is placed between the fog computing and cloud layer to synchronize all the incoming tasks based on the network traffic and resource availability. All the fog devices periodically update their available space to the SDN controller. As most of the tasks generated by the devices are latency sensitive, so the tasks are required to be executed in fog computing layer; however, there are other tasks that may be referred to as latency tolerant which are not hard time sensitive, so they can be executed in the cloud data center, if any fog cluster is overloaded. The SDN controller is regularly updated with all the information about current capacity/overload/underload. A typical fog control manager functional model is shown in Figure 7.8.

7.8 Mathematical Models

7.8.1 IoT Application Model

An IoT application can be considered as a finite state machine (FSM) represented as 5 tuple $A = (\Psi, \sum, \Theta, D, d, \mathbf{s})$ where Ψ represents the non-empty finite tasks, $\sum$ represents the input data from all the sensor devices and Θ represents the output of the tasks, D represents the dependency between the tasks, d represents the deadline assigned by the application developer, and $\mathbf{s}$ represents the occurrence of IoT application. If $D = 0$, then we say that IoT applications are independent tasks; otherwise they are said to be dependent.

Independent task model: An IoT application is modeled as a set of independent tasks represented as $A = T_i = \{t_i^1, t_i^2, t_i^3 \dots t_i^n\}$ where i and n, represent the indices of application and task respectively.

Dependent task model: A dependent task model is represented as a directed acyclic graph $G = (V, E)$ where V, represents the tasks and E represents the edges between the tasks.

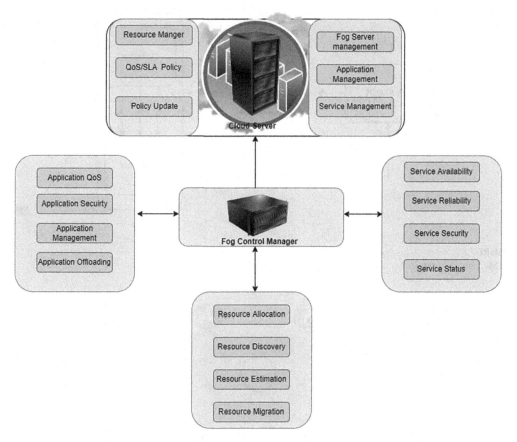

FIGURE 7.8
Fog control manager functions.

A task having no parent task is called an entry task whereas a task having no child node is called an exit task.

7.8.2 FCM Queueing Model

A single server queue M/M/1 is shown in Figure 7.9. It is used in the FCM to store the task before it is assigned to the appropriate fog node. Task arrivals follow a Poisson distribution with a first-come-first-served (FCFS) discipline. Hence, the FCM system and application request can be modeled as a birth–death process where the birth rate or arrival rate is λ and death rate or service rate is μ. It is also assumed that the arrival rate is less than the service rate, i.e. ($\lambda < \mu$). A birth and death model is shown in Figure 7.10.

(i) At time $t = 0$ the probability that the FCM is free can be determined using the steady-state probability formula given as

$$\text{Probability}\left(\text{FCM} = \text{empty}\right) = \frac{1}{1 + \sum_{k=1}^{\infty} \prod_{i=0}^{k-1} \frac{\lambda}{\mu}}$$

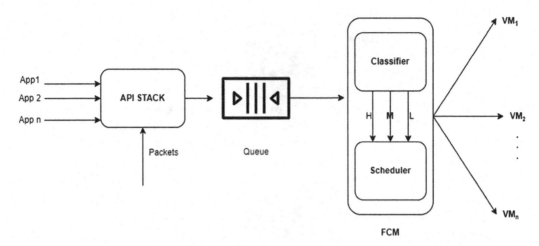

FIGURE 7.9
M/M/1 queue model.

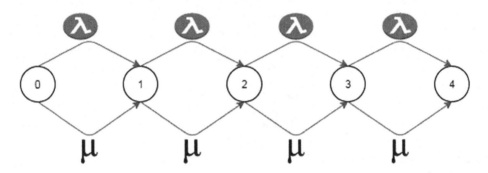

FIGURE 7.10
Birth and death model.

Where,

(ii) Probability of n tasks arrivals in time $(t + \delta t) = P_n(t) = \dfrac{(\lambda t)^n e^{-\delta t}}{n!}$ where mean and variance, both equals to (λt)

(iii) The mean response time of the M/M/1 queue is determined by using the equation $T = \dfrac{1/\mu}{1-\rho}$

(iv) The mean waiting time of an IoT task when the queue is busy can be determined by using the equation $W = \dfrac{p/\mu}{1-\rho}$

In order to provide scalability, the FCM node periodically (less than 1 sec) checks the application request frequency and the current status of the FCM. If the task arrival rate is higher than the service rate, the FCM creates a new instance of fog device to allocate the incoming tasks, whereas if the current status of the queue is empty, i.e., the arrival rate is less than the service rate, it will automatically delete the fog device. For efficient resource management of the edge layer, a threshold is fixed by the edge service provider to scale the edge resources as per the requirement.

7.9 Research Challenges

Despite the great advantages of fog computing and SDN for IoT applications, there are a number of challenges in edge computing arising from the emergence of various artificial intelligence- and machine learning-based IoT applications, enlisted as follows:

Task offloading decision: A multi-criterion decision problem when heterogeneity exists between server and clients. In the proposed architecture, both fog devices and requests from the respective IoT devices are heterogeneous, meaning the APIs of each IoT application may differ in terms of latency sensitivity and their objectives may also differ, for example, as regards deadlines. Various multi-criterion optimization systems are employed for effective offloading decisions. The offloading optimization problem is a genuine problem in edge computing, because most of the edge devices are computation constrained and energy constrained. Intelligent decisions to offload some tasks to the cloud based on task characteristics are therefore vital.

Service provisioning: This is another problem in the fog computing model. How the fog service provider provides resources to the different edge devices is again critical, because the communication technologies used by these edge devices are different. If certain services are not available, then migrating service from one edge device to another is a challenge.

Task Migration: Some IoT application tasks need certain resources that are not available in a single edge device, hence there is a need for task migration to other virtual machines, specifically in intelligent transport systems.

Resource estimation: The amount of resource required for a particular IoT application cannot be predicted before its execution.

7.10 Conclusion and Future Scope

It is concluded that most research works in edge computing deals with the optimization of conventional performance metrics like latency, energy and cost in order to maximize resource utilization. The importance of other qualitative parameters like scalability, reliability, security and availability in the context of the growing number of IoT applications has not been properly discussed. However, in this chapter, a FCM is proposed, that primarily aims to allocate the various IoT tasks to the available resources by selecting the appropriate module. The heterogeneous nature of IoT devices sometimes requires enormous resources to execute a particular IoT application. In order to maintain the ever-growing resource demand from IoT users, FCM is capable of scaling up the resources and, equally, in case of fewer requests it can easily scale down the resources by deleting virtual machines from the edge computing layer. The edge computing service provider must set a threshold value for the number of requests it can handle. If IoT requests cross the threshold value it automatically creates many instances of VM to maintain QoS. In the future, it is planned to apply the same procedure to advanced virtualization technologies such as container and serverless architecture to support the heterogeneous IoT requirements.

Acknowledgements: This work was supported by the Kempe post-doc fellowship via Project No. SMK21-0061, Sweden. Additional support was provided by the Wallenberg AI, Autonomous Systems and Software Program (WASP) funded by Knut and Alice Wallenberg Foundation.

References

[1] Chandak, A., Sahoo, B., & Turuk, A. K. (2011). An overview of task scheduling and performance metrics in grid computing.

[2] Apat, H. K., Sahoo Compt, B., Bhaisare, K., & Maiti, P. (2019, December). An optimal task scheduling towards minimized cost and response time in fog computing infrastructure. In *2019 International Conference on Information Technology (ICIT)* (pp. 160–165). IEEE.

[3] Tiwary, M., Puthal, D., Sahoo, K. S., Sahoo, B., & Yang, L. T. (2018). Response time optimization for cloudlets in mobile edge computing. *Journal of Parallel and Distributed Computing*, *119*, 81–91.

[4] Yousefpour, A., Patil, A., Ishigaki, G., Kim, I., Wang, X., Cankaya, H. C., ... & Jue, J. P. (2019). FOGPLAN: A lightweight QoS-aware dynamic fog service provisioning framework. *IEEE Internet of Things Journal*, *6*(3), 5080–5096.

[5] Singh, S. P., Nayyar, A., Kumar, R., & Sharma, A. (2019). Fog computing: from architecture to edge computing and big data processing. *The Journal of Supercomputing*, *75*(4), 2070–2105.

[6] Singh, S. P., Nayyar, A., Kaur, H., & Singla, A. (2019). Dynamic task scheduling using balanced VM allocation policy for fog computing platforms. *Scalable Computing: Practice and Experience*, *20*(2), 433–456.

[7] Barot, S. R., Goyal, S., & Kumar, A. (2019). Internet of Things: Inception and Its Expanding Horizon. In *Handbook of IoT and Big Data* (pp. 289–307). CRC Press.

[8] Shah, S. D. A., Gregory, M. A., Li, S., & Fontes, R. D. R. (2020). SDN enhanced multi-access edge computing (MEC) for E2E mobility and QoS management. *IEEE Access*, *8*, 77459–77469.

[9] Sun, X., & Ansari, N. (2016). EdgeIoT: Mobile edge computing for the Internet of Things. *IEEE Communications Magazine*, *54*(12), 22–29.

[10] Yang, F., Wang, S., Li, J., Liu, Z., & Sun, Q. (2014). An overview of internet of vehicles. *China Communications*, *11*(10), 1–15.

[11] Dandala, T. T., Krishnamurthy, V., & Alwan, R. (2017, January). Internet of vehicles (IoV) for traffic management. In *2017 International Conference on Computer, Communication and Signal Processing (ICCCSP)* (pp. 1–4). IEEE.

[12] Contreras-Castillo, J., Zeadally, S., & Guerrero-Ibañez, J. A. (2017). Internet of vehicles: architecture, protocols, and security. *IEEE Internet of Things Journal*, *5*(5), 3701–3709.

[13] Yang, F., Li, J., Lei, T., & Wang, S. (2017). Architecture and key technologies for Internet of Vehicles: a survey. *Journal of Communications and Information Networks*, *2*(2), 1–17.

[14] https://www.cisco.com/c/en/us/solutions/internet-of-things/future-of-iot.html

[15] https://www.idc.com/getdoc.jsp?containerId=prAP46737220

[16] Bonomi, F., Milito, R., Zhu, J., & Addepalli, S. (2012, August). Fog computing and its role in the internet of things. In *Proceedings of the First Edition of the MCC Workshop on Mobile Cloud Computing* (pp. 13–16).

[17] Skarlat, O., Nardelli, M., Schulte, S., & Dustdar, S. (2017, May). Towards qos-aware fog service placement. In *2017 IEEE 1st International Conference on Fog and Edge Computing (ICFEC)* (pp. 89–96). IEEE.

[18] Thota, C., Sundarasekar, R., Manogaran, G., Varatharajan, R., & Priyan, M. K. (2018). Centralized fog Computing Security Platform for IoT and Cloud in Healthcare System. In *Fog Computing: Breakthroughs in Research and Practice* (pp. 365–378). IGI Global.

[19] Sahoo, K. S., Mohanty, S., Tiwary, M., Mishra, B. K., & Sahoo, B. (2016, August). A comprehensive tutorial on software defined network: The driving force for the future internet technology. In *Proceedings of the International Conference on Advances in Information Communication Technology & Computing* (pp. 1–6).

[20] Rout, S., Sahoo, K. S., Patra, S. S., Sahoo, B., & Puthal, D. (2021). Energy efficiency in software defined networking: A survey. *SN Computer Science, 2*(4), 1–15.

[21] Mohanty, S., Kanodia, K., Sahoo, B., & Kurroliya, K. (2020, February). A simulated annealing strategy for reliable controller placement in software defined networks. In *2020 7th International Conference on Signal Processing and Integrated Networks (SPIN)* (pp. 844–849). IEEE.

[22] Toosi, A. N., Son, J., & Buyya, R. (2018). Clouds-pi: A low-cost raspberry-pi based testbed for software-defined-networking in cloud data centers. *ACM SIGCOMM Comput Commun Review, 7*, 1–11.

[23] Kaur, A., Singh, P., & Nayyar, A. (2020). Fog Computing: Building a Road to IoT with Fog Analytics. In *Fog Data Analytics for IoT Applications* (pp. 59–78). Springer, Singapore.

[24] Singh, S. P., Kumar, R., Sharma, A., & Nayyar, A. (2020). Leveraging energy-efficient load balancing algorithms in fog computing. *Concurrency and Computation: Practice and Experience,* e5913.

[25] Nayyar, A. (2019). *Handbook of Cloud Computing: Basic to Advance Research on the Concepts and Design of Cloud Computing.* BPB Publications.

[26] Sahoo, S., Pattanayak, A., Sahoo, K. S., Sahoo, B., & Turuk, A. K. (2018, December). MCSA: A multi-constraint scheduling algorithm for real-time task in virtualized cloud. In *2018 15th IEEE India Council International Conference (INDICON)* (pp. 1–6). IEEE.

[27] Sahoo, S., Sahoo, K. S., Sahoo, B., & Gandomi, A. H. (2020, December). An auction based edge resource allocation mechanism for IoT-enabled smart cities. In *2020 IEEE Symposium Series on Computational Intelligence (SSCI)* (pp. 1280–1286). IEEE.

[28] Sahoo, K. S., Tiwary, M., Luhach, A. K., Nayyar, A., Choo, K. K. R., & Bilal, M. (2021). Demand-supply based economic model for resource provisioning in industrial IoT traffic. *IEEE Internet of Things Journal.*

[29] Ghobaei-Arani, M., Souri, A., & Rahmanian, A. A. (2020). Resource management approaches in fog computing: a comprehensive review. *Journal of Grid Computing, 18*(1), 1–42.

[30] Mishra, S. K., Mishra, S., Alsayat, A., Jhanjhi, N. Z., Humayun, M., Sahoo, K. S., & Luhach, A. K. (2020). Energy-aware task allocation for multi-cloud networks. *IEEE Access, 8*, 178825–178834.

[31] Sahu, S. K., Mohapatra, D. P., Rout, J. K., Sahoo, K. S., Pham, Q. V., & Dao, N. N. (2022). A LSTM-FCNN based multi-class intrusion detection using scalable framework. *Computers and Electrical Engineering, 99*, 107720.

[32] Maity, P., Saxena, S., Srivastava, S., Sahoo, K. S., Pradhan, A. K., & Kumar, N. (2021). An effective probabilistic technique for DDoS detection in OpenFlow controller. *IEEE Systems Journal.*

[33] Sharma, P. K., Rathore, S., Jeong, Y. S., & Park, J. H. (2018). SoftEdgeNet: SDN based energy-efficient distributed network architecture for edge computing. *IEEE Communications Magazine, 56*(12), 104–111.

[34] Jararweh, Y., Otoum, S., & Al Ridhawi, I. (2020). Trustworthy and sustainable smart city services at the edge. *Sustainable Cities and Society, 62*, 102394.

[35] Pekar, A., Mocnej, J., Seah, W. K., & Zolotova, I. (2020). Application domain-based overview of IoT network traffic characteristics. *ACM Computing Surveys (CSUR), 53*(4), 1–33.

[36] Gu, L., Zeng, D., Guo, S., Barnawi, A., & Xiang, Y. (2015). Cost efficient resource management in fog computing supported medical cyber-physical system. *IEEE Transactions on Emerging Topics in Computing, 5*(1), 108–119.

[37] Minh, Q. T., Tran, C. M., Le, T. A., Nguyen, B. T., Tran, T. M., & Balan, R. K. (2018, October). Fogfly: A traffic light optimization solution based on fog computing. In *Proceedings of the 2018 ACM International Joint Conference and 2018 International Symposium on Pervasive and Ubiquitous Computing and Wearable Computers* (pp. 1130–1139).

[38] Pallewatta, S., Kostakos, V., & Buyya, R. (2019, December). Microservices-based IoT application placement within heterogeneous and resource constrained fog computing environments. In *Proceedings of the 12th IEEE/ACM International Conference on Utility and Cloud Computing* (pp. 71–81).

[39] Sharif, A., Nickray, M., & Shahidinejad, A. (2020). Fault-tolerant with load balancing scheduling in a fog-based IoT application. *IET Communications, 14*(16), 2646–2657.

[40] Xia, J., Cheng, G., Guo, D., & Zhou, X. (2020). A QoE-aware service-enhancement strategy for edge artificial intelligence applications. *IEEE Internet of Things Journal, 7*(10), 9494–9506.

[41] Farahani, B., Firouzi, F., Chang, V., Badaroglu, M., Constant, N., & Mankodiya, K. (2018). Towards fog-driven IoT eHealth: Promises and challenges of IoT in medicine and healthcare. *Future Generation Computer Systems, 78*, 659–676.

[42] Korala, H., Jayaraman, P. P., Yavari, A., & Georgakopoulos, D. (2020, November). APOLLO: A platform for experimental analysis of time sensitive multimedia IoT applications. In *Proceedings of the 18th International Conference on Advances in Mobile Computing & Multimedia* (pp. 104–113).

[43] Venticinque, S., & Amato, A. (2019). A methodology for deployment of IoT application in fog. *Journal of Ambient Intelligence and Humanized Computing, 10*(5), 1955–1976.

[44] Teng, H., Liu, Y., Liu, A., Xiong, N. N., Cai, Z., Wang, T., & Liu, X. (2019). A novel code data dissemination scheme for Internet of Things through mobile vehicle of smart cities. *Future Generation Computer Systems, 94*, 351–367.

[45] Arora, S., & Bala, A. (2020). Pap: power aware prediction based framework to reduce disk energy consumption. *Cluster Computing, 23*(4), 3157–3174.

[46] Ahmed, E., Yaqoob, I., Hashem, I. A. T., Khan, I., Ahmed, A. I. A., Imran, M., & Vasilakos, A. V. (2017). The role of big data analytics in Internet of Things. *Computer Networks, 129*, 459–471.

[47] Ur Rehman, M. H., Yaqoob, I., Salah, K., Imran, M., Jayaraman, P. P., & Perera, C. (2019). The role of big data analytics in industrial Internet of Things. *Future Generation Computer Systems, 99*, 247–259.

[48] Wang, L., Von Laszewski, G., Younge, A., He, X., Kunze, M., Tao, J., & Fu, C. (2010). Cloud computing: a perspective study. *New Generation Computing, 28*(2), 137–146.

[49] Mell, P., & Grance, T. (2011). *The NIST Definition of Cloud Computing.*

[50] Kim, W. (2009). Cloud computing: Today and tomorrow. *The Journal of Object Technology, 8*(1), 65–72.

[51] Abbas, N., Zhang, Y., Taherkordi, A., & Skeie, T. (2017). Mobile edge computing: A survey. *IEEE Internet of Things Journal, 5*(1), 450–465.

[52] Mao, Y., You, C., Zhang, J., Huang, K., & Letaief, K. B. (2017). A survey on mobile edge computing: The communication perspective. *IEEE Communication Surveys and Tutorials, 19*(4), 2322–2358.

[53] Sun, X., & Ansari, N. (2016). EdgeIoT: Mobile edge computing for the Internet of Things. *IEEE Communications Magazine, 54*(12), 22–29.

[54] Yu, W., Liang, F., He, X., Hatcher, W. G., Lu, C., Lin, J., & Yang, X. (2017). A survey on the edge computing for the Internet of Things. *IEEE Access, 6*, 6900–6919.

[55] Chen, B., Wan, J., Celesti, A., Li, D., Abbas, H., & Zhang, Q. (2018). Edge computing in IoT-based manufacturing. *IEEE Communications Magazine, 56*(9), 103–109.

[56] Krishnamurthi, R., Kumar, A., Gopinathan, D., Nayyar, A., & Qureshi, B. (2020). An overview of IoT sensor data processing, fusion, and analysis techniques. *Sensors, 20*(21), 6076.

[57] Rahman, L. F., Ozcelebi, T., & Lukkien, J. (2018). Understanding IoT systems: a life cycle approach. *Procedia Computer Science, 130*, 1057–1062.

[58] https://www.mckinsey.com/~/media/McKinsey/Industries/Technology

[59] Dastjerdi, A. V., & Buyya, R. (2016). Fog computing: Helping the internet of things realize its potential. *Computer, 49*(8), 112–116.

[60] Salman, O., Elhajj, I., Chehab, A., & Kayssi, A. (2018). IoT survey: An SDN and fog computing perspective. *Computer Networks, 143*, 221–246.

8

Deep Learning Models with SDN and IoT for Intelligent Healthcare

Vedaant Singh and Srishty Singh
KIIT University, India

CONTENTS

DOI: 10.1201/9781003213871-8

8.1 Introduction

There have been ongoing usability and usefulness obstacles to the utilization of digital innovation in healthcare. The fusion of numerous healthcare paradigms has been sluggish, and the majority of the world's countries have yet to adopt a fully engaged healthcare system. The study of human biology and patients' individual variances have continually established the importance of the human factor in disease, treatment and diagnosis. Advances in technology, on the other hand, are undeniably becoming indispensable tools for healthcare providers in providing the best possible care to their patients.

The internet of things (IoT) is a key technology that allows billions of items and smart devices to connect to the internet and operate with little or no user involvement. Heuristically, such a massive network of intelligent networked devices generates a lot of traffic, which makes it more difficult to handle with standard network architecture. Because of the heterogeneity of IoT devices, popular communication technologies and protocols [1, 2] are required, as well as the ability to work efficiently with low-power devices in a chaotic and uncompressed communication context.

The prevalence of heterogeneous devices that create massive bandwidth is one of the fundamental and unavoidable difficulties in IoT. Controlling such diverse devices with high traffic loads is still an unsolved research problem. Fortunately, a recent focus area known as software-defined network (SDN) offers advantages including centralized network management and programming [3].

Thanks to developments in computing technology such as storage systems, computational resources, and data transmission velocities, machine learning has become widely used in a variety of sectors, particularly healthcare. Modern medical advances have emphasized the need for a customized medication or "precision medicine" strategy, because of the multifaceted nature of giving effective healthcare. The goal of personalized medicine is to use massive amounts of health records to reveal, anticipate, and analyse diagnostic decisions that clinicians can subsequently use for a specific person.

The goal is to concentrate on three of the most significant machine learning (ML) applications in the bio-pharmaceutical area. Deep learning offers a wide range of potential applications in the healthcare industry, which may cover auxiliary elements of the industry, namely staff management, technical services, insurance policies, and many more, since it is a rapidly increasing sector. As a result, the topics covered in this chapter are limited to three common machine learning applications. The most pertinent one is the application of deep learning to medical imaging such as magnetic resonance imaging (MRI), computerized axial tomography (CAT) scans, positron emission tomography (PET) scans and ultrasound imaging. The ultimate result of these imaging procedures is a collection or series of images, which must be analyzed and diagnosed by a radiologist. Machine learning techniques have made significant progress in their ability to anticipate and identify pictures that could indicate a pathological condition or a real concern.

SDNs can also help healthcare firms manage many linked device deployments. The growing use of IoT devices necessitates the use of a flexible infrastructure. The internet of things (IoT) provides a seamless platform for human-to-human interactions with a variety of physical and virtual objects, enabling healthcare monitoring domains [4]. SDNs aid enterprises in gathering, sending, and processing data by allowing them to better manage and monitor physical equipment. That information can be used to improve IoT applications. The more efficiently a network can handle and send IoT data, the more useful it is for developing apps. Many medical IoT devices can store some patient data, however when signal changes a require the device to alert a clinician, significant and prioritized communication is needed. For example, if a heart monitor identifies a life-threatening heart anomaly, that information should be sent to a professional first. Telehealth programs are also supported by SDNs. In this context, the combination of SDN with the internet of multimedia things (IoMT), known as SD-IoMT, offers enormous potential to increase the IoT system's network management and security capabilities [5].

Another is natural language processing in medical records. Several doctors agree that the transition to electronic medical records (EMR) in many countries is slow, tedious, and, in many circumstances, completely mismanaged. Patients may receive much poorer treatment as a result of this. One of the most serious difficulties is the large amount of tangible medical information and documentation that already exists in numerous healthcare facilities. Due to differences in formatting, written entries, and a variety of incomplete or noncentralized data, the move to electronic medical records has been gradual.

Organization of the Chapter

Section 8.2 elaborates related work. Section 8.3 describes the methodology. Section 8.4 highlights case studies. Section 8.5 discusses simulation results and analysis. Section 8.6 concludes the chapter with future scope.

8.2 Related Work

The use of deep learning and machine learning for medical purposes is relatively new, and it has yet to be thoroughly explored. The sections that follow cover some of the most recent research on the use of deep models in clinical imaging, electronic health records (EHRs), wearable device data, and genomics. Convolutional neural networks (CNNs) [6], recurrent neural networks (RNNs) [7], restricted Boltzmann machines (RBMs) [8], and autoencoders (AEs) [9] are the most common deep learning architectures used in the medical services field.

Lieu et al. [10] suggested a layered sparsed AE for early Alzheimer's disease detection from brain MRIs. Esteva et al. [11] suggested a CNN model for skin cancer categorization at the dermatologist level. Toh and Brody [12] discussed how deep learning and machine learning are changing the healthcare industry around the world. Dernoncourt et al. [13] suggested an LSTM RNN for patient clinical note de-identification. To predict chromatin marks from DNA sequences, Zhou et al. [14] proposed a CNN architecture. Kelley et al. [15] used CNN to create Basset, an open-source platform for predicting DNase I hypersensitivity in several cell types and quantifying the impact of SNVs on chromatin accessibility. Prasoon et al. [16] proposed a CNN model for segmenting knee cartilage MRIs automatically to predict the risk of osteoarthritis. Liu et al. [17] developed a stacked denoising AE for using ultrasound images to diagnose breast nodules and lesions. Nguyen et al. [18]

presented an end-to-end CNN for Deeper to predict unplanned readmission after discharge. Choi et al. [19] employed GRU RNN to create a Doctor AI that leverages a patient's history to anticipate diagnosis and prescriptions for a future visit. CNN was utilized by Koh et al. [20] to determine the prevalence of various chromatin markers. Healthcare 4.0 processes for accessing data, according to Kumar et al. [21], should be tested using statistical simulation optimization methodologies and algorithms. Authors also provides a thorough and comparable analysis of current blockchain-based smart healthcare solutions.

The need for remote healthcare was re-established by Pramanik et al. [22]. Telemedicine and remote monitoring are the two techniques of remote healthcare that are explored. The issue of security and privacy in remote healthcare was their main point of interest. Pramanik et al. [23] set out to investigate the clinical and medical implications of various nanotechnology implementations. They provided a thorough introduction to nanotechnology, biosensors, nanobiosensors, and the internet of things. Nanotechnology, nanoparticles, biosensors, nanobiosensors, and nanozymes were given as multilayer taxonomies. To improve the fault detection rate, Ali et al. [24] proposed a test-case selection and prioritizing framework based on a design pattern. With reference to electronic healthcare records, Mahapatra et al. [25] suggested several methods to discuss three primary techniques, including data storage, data access, and data authentication, in order to keep patient records safe and adhere to the confidentiality, integrity, and availability concept. The principles of the state-of-the-art machine learning-based system for healthcare were explained by Nayyar et al. in [26]. R. Krishnamurthi et al. [27] presented different fog-based data processing and data analysis (FDPA) techniques in fog computing solutions to achieve Healthcare 4.0 goals. Nayyar et al. [28] presented cutting-edge AR and VR technology for tourism and hospitality. In comparison to existing encryption approaches, Azeez et al. [29] introduced an efficient affine cipher-based encryption methodology that provides a high level of confidentiality with a smaller key size. In the security analysis part, the suggested work's security strength against various harmful security threats is demonstrated to ensure that it delivers greater security. Alzubi et al. [30] presented an introduction to data analytics methodology, which allows machines to understand and perform what humans do instinctively, namely learn from experiences. It covers the fundamentals of machine learning, including definitions, nomenclature, and implementations that illustrates the what, how, and why it works. Kumar et al. [31] provided an in-depth and well-organized overview of multimedia social big-data mining and applications.

In this work, we offer an in-depth look at how deep learning has impacted the healthcare system around the world, as well as a comparison and analysis of the performance of various deep learning models.

8.3 Methodology

The following sections briefly describe the concept behind SDN, deep learning and the IoT. Two case studies of deep learning and IoT are also illustrated. Figure 8.1 illustrates the flow of the entire deep learning process.

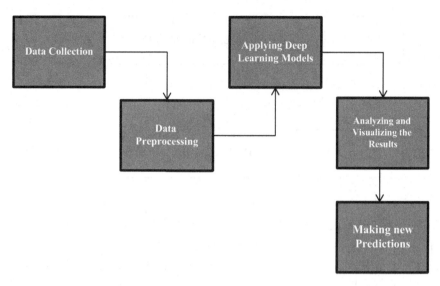

FIGURE 8.1
Flowchart of the entire deep learning process.

8.3.1 SDN Overview

SDN (software-defined networking) is a modern computing network paradigm that separates the data plane from the control plane to simplify network services. Unlike traditional architectures, which combine the control plane and the data plane in the same piece of equipment, the goal of SDN is to separate the data plane from the control plane to simplify network services [32].

The data plane is dispersed, but the control plane is centralized in SDN. The network's flexibility and flow-forward decision-making process are aided by the control plane's centralized design. The data plane is dispersed, but the control plane is centralized in SDN. The network's flexibility and flow-forward decision-making process are aided by the control plane's centralized design. SDN controller is located in the control plane of a programmable SDN network. By configuring the SDN controller without replacing any equipment, specific regulations can be added to the architecture [33, 34].

The SDN architecture is distinguished by two types of interface: southbound and northbound. The southbound interface is in charge of establishing the set of forwarding device commands. The data transmission between the control plane elements and forwarding devices is also defined by the southbound interface. An API for developing apps is provided through the northbound interface. The southbound interface utilizes this interface to encapsulate low-level commands or software that are used to program forwarding devices. Figure 8.2 depicts these interfaces.

8.3.2 SDN for IoT

Internet of Things (IoT) connects an enormous number of gadgets to the internet. With standard system design, maintaining such a large quantity is either impossible or unviable [33]. The overall efficiency of IoT networks, which produce a large volume of data, can be aided by distributed systems. To govern and administer such connections, new procedures

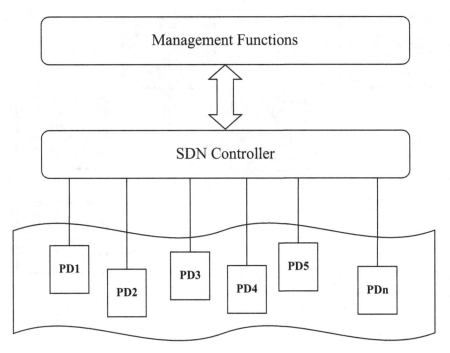

FIGURE 8.2
Software-defined networking architecture.

and structures are required. Using a global perspective, the SDN centralised functionality may be used to administer the network. As a result, enhancing bandwidth usage, load balancing, and decreasing delay can all help an IoT network perform better.

8.3.3 Approaches Used in Smart Healthcare

8.3.3.1 Machine Learning-Based Approach

This is the science of getting machines to learn without being explicitly programmed. It can be used very effectively in sentiment analysis. In general, there are two types of approaches based on the concept of machine learning: supervised and unsupervised learning.

Supervised learning: The most common type of training data used in deep classification is labeled data. In training data, there are numerous inputs and a "labeled" output. These labeled outcomes are used by models to assess themselves during training in order to enhance their capacity to predict new data [35–37]. Typically, supervised learning models focus on regression and classification methods. Categorization problems are fairly common in medicine. A doctor diagnoses a patient in most clinical settings by identifying the condition based on the symptoms. The purpose of regression problems is to predict quantitative outcomes, such as the length of a patient's visit to a clinic, using a group of variables including vital signs, clinical records, and weight. Some of the most common algorithms, under supervised learning are linear regression, logistic regression, decision tree, random forest, support vector machine. Even neural networks can be trained via supervised learning [38].

Unsupervised Learning. Here, the data has no labels and it self-organizes the model to find pattern in the dataset. Unsupervised learning looks for patterns in data that have not been tagged [39]. Such algorithms are usually good at grouping data into useful groups, enabling identification of latent traits that are not readily evident. These are, generally, more technically expensive and take a greater quantity of data to complete.

K-means clustering and deep learning are well-known and commonly used algorithms, with deep learning being supervised as well [40]. These algorithms also handle tasks linked to association, that are identical to grouping. These algorithms are referred to as unsupervised because no human interaction or interference is supplied as to which collection of characteristics the groups will focus on. Figure 8.3 shows the various branches of the machine learning strata.

8.3.3.2 Deep Learning-Based Approach

Deep learning is a branch of AI that is essentially a three-layered neural structure. This type of neural framework tries to mimic the human cerebrum's activity by allowing it to "learn" from massive amounts of raw data, but falls well short of its capabilities [41]. While a single-layer neural architecture may provide shaky predictions, adding more hidden layers and units could significantly increase accuracy. Figure 8.4 depicts a general neural network model.

8.3.3.3 Hyper-parameters

In deep learning, a model usually has a number of criteria and hyper-parameters. Model variables which can be altered during training are known as parameters. Criteria may include the amount of training data as well as the fact that each batch of data differs on one or more parameters. Hyper-parameters, on the other hand, are normally set prior to

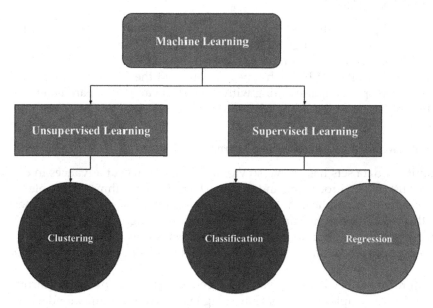

FIGURE 8.3
Machine learning structure/ Branches of Machine Learning.

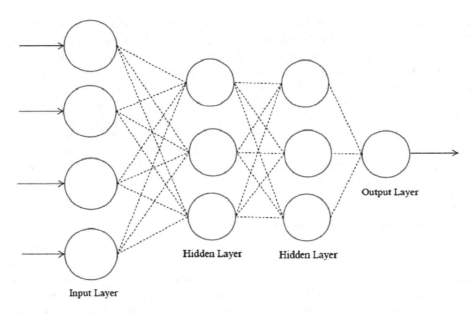

FIGURE 8.4
A neural network architecture.

training and cannot be modified once the learning process has begun. Hyper-parameters are frequently used to fine-tune features like the model's learning rate, and to limit the algorithm. Many researchers [42, 43] have attempted to make hyper-parameter tuning more efficient and manageable.

8.3.3.4 Algorithm Principles

Given the rapid pace of studies in this area, most of these deep learning approaches are always evolving and improving, but it is vital to realize that not all algorithms are suitable for all use situations. Each algorithm has its own set of benefits and drawbacks. Specific algorithms may be affected by such types of data, and the hours spent developing such models are often spent experimenting with numerous variations, parameters and hyper-parameters in these techniques in order to obtain the best generic outcome.

8.3.4 Smart Healthcare Using Machine Learning

The quantity of data sets has grown so vast as a result of recent advances in digital technology that usual data processing and machine learning algorithms are unable to deal efficiently. Interpreting complicated, high-dimensional, and noise-contaminated sets of data, on the other hand, is a significant problem, and unique algorithms that can consolidate, classify, extract vital information, and transform them to a comprehensible form are critical. Deep learning models have shown exceptional results in tackling these difficulties in the last decade.

The future of artificial intelligence has been transformed by deep learning. It has resolved a slew of complex issues that have plagued the AI community for decades. Models

TABLE 8.1

Different Architectures (Deep Learning)

Architecture	Description
Convolutional neural network (CNN)	Deep learning model that can take an image and assign significance (learnable weights and biases) to distinct perspectives/objects in the image, as well as separate them.
Recurrent neural network (RNN)	Sort of artificial neural network in which nodes are linked in a chart that follows a chronological order.
Restricted Boltzmann machine (RBM)	Generative probabilistic fake neural structure that is capable of learning a likelihood appropriation over its bits of feedbacks.
Autoencoder (AE)	Type of unsupervised artificial neural network that learns efficient data codings.

related to DL are, in fact, more advanced versions of artificial neural networks (ANNs) with numerous layers, presumably linear or non-linear. Every layer has distinct loads that link it to the bottom and higher layers. DL models' capacity to learn hierarchical features from a variety of data formats, including as mathematical, visual, textual and sound, makes them more effective for solving detection, regression, semi-supervised, and unsupervised issues.

Many deep architectures along with various learning paradigms have been rapidly observed in recent years to develop machines that can perform as well as or better than humans in a variety of domains, including clinical diagnosis, self-driving automobiles, language processing, image processing and probabilistic forecasting. Table 8.1 reviews the different neural network organizations modeling the deep learning designs applicable to the medical care domain.

8.4 Case Studies

8.4.1 Case I: Study of Skin Cancer Using Deep Learning (CNN)

The deep learning model developed for analysis of skin cancer data involves various steps as shown in Figure 8.1.

8.4.1.1 Dataset

The assortment of biological data making up these data sets shows no sign of slowing down. These days data can easily be gathered or grasped with the help of various online sources such as Kaggle or other publicly available databases. The ISIC (International Skin Image Collaboration) archive provided the dataset. It contains approximately 1800 images of generous (benign) moles and 1497 images of harmful (malignant) moles. All the photographs have been reduced in size to 224x224x3 RGB. The model will be evaluated based on the accuracy score, (TP + TN) / (TOTAL). Figures 8.5 and 8.6 show the two types of skin cancer: benign (non-cancerous) and malignant (cancerous).

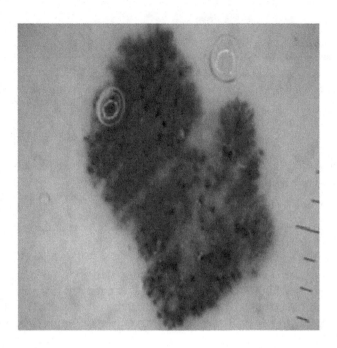

FIGURE 8.5
A malignant skin cancer.

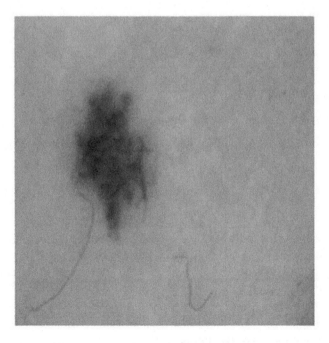

FIGURE 8.6
A benign skin cancer.

8.4.1.2 CNN Model and Architecture

The CNN (or convolutional neural network) is a deep learning version of the ANN (or artificial neural network), with every layer learning from the input layers of neurons (rather than just adjusting the weights as is the case with ANN). CNNs have been widely employed in a range of applications, such as traffic prediction and picture classification, and due to their capacity to learn hidden features [44–46].

An input layer, an output layer, and numerous hidden layers in between make up the basic architecture of a CNN model for cancer diagnosis. The CNN model's hidden layers also include several rounds of convolutional and pooling layers, which are then coupled to the fully connected layers before producing an output.

8.4.2 Case II: How Deep Learning and IoT Help Make Healthcare Effective

Deep learning is gradually making inroads into cutting-edge tools with rising numbers of applications in the medical environment. Modern patient-facing applications, and also some very surprisingly well-established tactics for enhancing health IT user experience, are among the most viable use cases.

Diagnostics and imaging analytics: CNN is a kind of deep learning model that is essentially well-suited to processing images such as MRI or X-ray data. According to computer programming researchers at Stanford University, CNN is built on the premise that it will process images, permitting the models to run more effectively and process larger images.

Natural Language Processing Systems: Several natural language processing systems that have grown prominent in the healthcare sector for drafting paperwork and converting speech to text have used deep learning and neural networks. Because neural networks are built for categorization, they may recognize specific language or grammatical features by "grouping" related words and projecting them to each other. This aids in the network's comprehension of complicated semantic meaning. However, the intricacies of everyday speech and conversation make the task more difficult. While acceptable speech-to-text accuracy has become a fairly standard capability for narration systems, extracting valuable and consistent observations via free-text medical data is far more difficult.

Precision medicine, drug discovery, predictive analysis, clinical decision assistance: Unexpected findings are common in the realm of genetic medicine, making it an attractive testing ground for novel approaches to focused care. The National Cancer Institute and the Department of Energy are fostering a culture of exploration through a number of collaborative projects focusing on harnessing machine learning for cancer research. Using a mixture of predictive analytic and molecular modeling, researchers hope to learn more about how and why particular malignancies develop in certain people. Deep learning technology, according to the two organizations, will speed up data analysis by reducing processing time for essential parts from weeks to months to just a few moments. In a similar manner, the pharmaceutical industry has great aspirations for the involvement of deep learning in clinical decision support and predictive analytics for a huge range of circumstances. Deep learning could potentially be a valuable tool in hospital settings, alerting doctors to changes in increased risks in illnesses like sepsis and respiratory failure (see Figure 8.7).

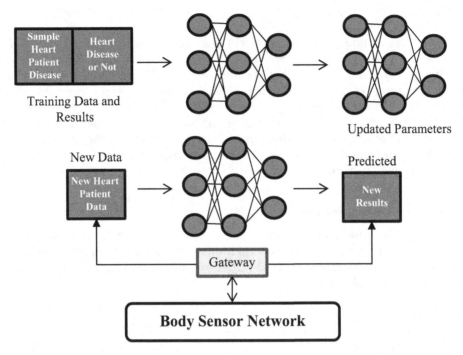

FIGURE 8.7
An ensemble deep learning-based smart healthcare system.

8.5 Simulation Results and Discussion

Skin cancer is one of the most common cancers in humans, and it is evaluated primarily from the outside in, beginning with an underlying clinical screening, maybe followed by dermascopic examination, a biopsy, and histological evaluation. Because of the fine-grained heterogeneity in the appearance of skin lesions, automated classification of skin lesions using photographs is a difficult task. To begin with, we did perform various exploratory data analysis before performing the cancer classification. This analysis gave us a better intuition and understanding of the two different scenarios related to the dataset – malignant and benign. Figure 8.8 depicts the accuracy comparison between the training and test set achieved by the custom CNN. Figure 8.9 provides a comparison between losses of the training and test sets. Similarly, Figure 8.10 depicts the accuracy and loss comparison between the training and test set using the VGG16 model. VGG16 outperformed the custom CNN, showing promising results.

The performance of classification models has been evaluated using four indices which are calculated as follows:

- Precision = TP/(TP+FP)
- Accuracy = (TP+TN)/(TP+TN+FP+FN)
- Recall = TP/(TP+FN)
- F1-score = (2×Precision×Recall)/ (Precision + Recall)
- Confusion Matrix

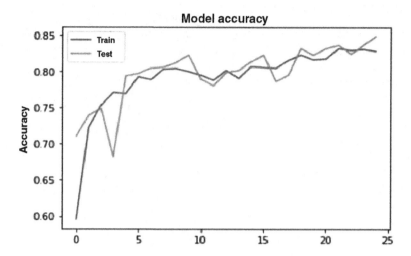

FIGURE 8.8
Training vs. test accuracy (custom CNN).

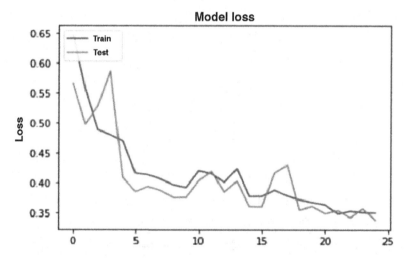

FIGURE 8.9
Training vs. test loss (custom CNN).

where TP stands for true positive count, TN stands for true negative count, FP stands for false positive count and FN refers to the false negative count, respectively. The dataset was split into two parts, i.e., training and test sets, where 75% of the total dataset was taken into consideration as the training set and the remaining 25% was considered as the test set. Table 8.2. shows the accuracy levels achieved using different models.

Finally, Table 8.3 compares the approaches. In this chapter, we have implemented different neural network models and enhanced the accuracy to 88.47%.

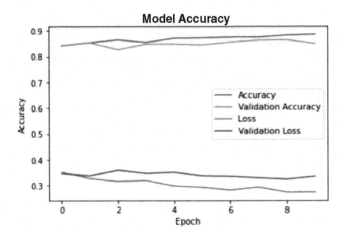

FIGURE 8.10
Training vs. test accuracy and loss (VGG 16 Model).

TABLE 8.2

Training and Test Data

Training Data	2637
Malignant	1197
Benign	1440
Test Data	660
Malignant	300
Benign	360

TABLE 8.3

Benchmarking of Our Approach

Sl. No.	Model	Accuracy (in %)
1.	Custom CNN	84.85
2.	VGG-16 [46]	88.47

8.5.1 Augmented IoT

With IoT-based healthcare technology, it is feasible to integrate multiple systems and establish interoperable services [47–50]. IoT applications in healthcare include tracking/monitoring, identification/authentication, sensing, and data collection/sharing. Data from many sources is coupled with intelligent decision-making and reasoning algorithms, enabling rapid data processing and complex analysis. Although online healthcare systems are still in their infancy, their market share is rapidly growing. Continuous health monitoring devices and distributed health management, which includes telemedicine, are some of the outcomes of the IoT approach.

8.5.2 Applications of Deep Learning and IoT

Real-life deep learning applications are a part of everyday life, but they are usually so effectively integrated into services and products that users are oblivious to the intricate data processing going on at the moment.

8.5.2.1 Drug Research and Production

ML and DL applications have made their way int the process of drug discovery, particularly in the basic stages, from the screening of ingredients to estimating a drug's rate of success based on biological parameters. The foundation for this is upcoming sequencing. Pharmaceutical businesses employ machine learning in their drug research and production process. However, at the moment, this is confined to unsupervised machine learning that can detect patterns in original data. The goal is to produce accurate medicine on the basis of unsupervised learning, which will allow doctors to discover pathways for "multifactorial" disorders.

8.5.2.2 Diagnosis and Identification of Diseases

Machine learning, in combination with deep learning, has aided significant advances in the diagnostic process. Doctors may now diagnose ailments which were formerly undiagnosable, such as early-stage tumors or cancers, as well as hereditary ailments, due to these modern tools. Robots allow surgeons to manipulate and control robotic arms to do precision surgery in restricted areas of the human body with minimal vibrations. Robotic surgery is often utilized in hair transplantation which requires exquisite precision and fragmentation. In the realm of surgery, robotics is currently leading the way. By incorporating real-time surgical metrics, insights from successful surgical experiences, and statistics from pre-op medical records into the surgical operation, robotics driven by AI and ML algorithms increase the efficacy of surgical equipment.

8.5.2.3 Personalized Treatment

ML technology can help build tailored treatments and drugs that specifically target diseases in individual patients by utilizing patient medical records. When integrated with predictive modeling, personalized treatment yields even more benefits. Instead of picking from a list of diseases or assessing the patient's risk depending on their clinical history, clinicians can use machine learning to diagnose their patients.

8.5.2.4 Clinical Trials

Deep learning technologies have a lot of potential when it comes to clinical trial research. Medical practitioners may analyze a broader range of information by using advanced predictive analytics that minimize the time and expenses required to conduct medical experimentation. According to McKinsey, a variety of machine learning applications can help improve clinical research efficiency, such as assisting in the selection of optimal sample sizes for enhanced efficacy and reducing the risk of data errors through the use of electronic health records (EHRs).

8.5.2.5 Enhanced Radiotherapy

In the discipline of radiology, deep learning has proven to be extremely beneficial. There are numerous discrete parameters in medical picture analysis that might be triggered at any time. Algorithms based on deep learning are useful in this case. Because deep learning algorithms adapt from a variety of data samples, they are better equipped to diagnose and detect the relevant variables. For example, in medical image analysis, ML is used to categorize items such as diseases into distinct categories such as normal or aberrant.

8.5.2.6 Maintenance of Health Records

It is common knowledge that managing and updating healthcare data and patient medical histories is a time-consuming and costly operation. DL technologies are assisting in the resolution of this problem by minimizing the amount of time, labor, and money spent on record keeping. Sorting and classifying healthcare data is made easier with the use of VMs (vector machines) and machine learning-based OCR recognition algorithms like Google's Cloud Vision API. There are smart health records, which allow doctors, health professionals, and patients to connect for the benefit of research, patient care, and public health.

8.5.2.7 Data Collection

Today's healthcare industry is heavily engaged in crowdsourcing healthcare data from a variety of sources (mobile applications, medical platforms, and so on), but only with people's agreement. Doctors and medical professionals can provide patients with timely and appropriate therapy based on this collection of real-time health data.

8.5.2.8 Epidemic Outbreak Prediction

Healthcare organizations use machine learning and artificial intelligence algorithms to track and anticipate potential epidemic outbreaks around the world. These digital systems can forecast outbreaks of disease by gathering satellite data, actual updates on social media, and other crucial information from the internet. This is especially beneficial for third-world nations that lack adequate healthcare facilities. Artificial neural networks and support vector machines, for example, have aided in the prediction of malaria outbreaks by taking into account characteristics such as temperature, average monthly precipitation, and so on. An internet-based software called ProMED-mail enables healthcare organizations to track diseases and anticipate epidemics in real time. HealthMap primarily depends on ProMED to monitor and inform countries about suspected epidemic outbreaks via automated categorization and visualization.

8.5.2.9 Analytics for Pattern Imaging

Machine learning techniques and algorithms are currently being used by healthcare organizations all over the world to improve image analytics and diagnostics. Machine learning technologies can help radiologists notice subtle changes in scans, helping them to recognize and treat health complications early.

8.5.3 Challenges

Despite the optimistic findings gained by using deep learning architectures, the practical applications of artificial intelligence to healthcare still have many unsolved questions.

8.5.3.1 Data Volume

Deep learning is an extremely expensive category of computer modeling. Fully linked multi-layer neural networks are a good example because they require an exact estimate of a large set of network properties. The availability of massive volumes of data is critical to achieving this goal. While there are no hard and fast rules about the minimal amount of training data, having at least ten times the amount of data as parameter estimates is a solid assumption. This is one of the reasons why deep learning is particularly useful in areas where massive amounts of data can be acquired quickly. Nevertheless, healthcare is a distinct domain; the world's population is only 7.9 billion people (as of June 2021), with the vast majority of them lacking access to quality healthcare. As a corollary, we will be unable to formulate a complete deep learning model with as many patients as we would want. Additionally, ailments and their variants are far more difficult to grasp than other tasks such as picture or voice recognition. As a result, the quantity of healthcare data required for training an efficient and successful deep learning model is significantly than for other media.

8.5.3.2 Data Quality

In comparison to other fields, where the data is well structured and accessible, medical research is incredibly diverse, puzzling, messy, and unfinished. It is challenging to train an efficient deep learning model with such huge and varied data sets due to difficulties like data sparsity, duplication, and irrelevant data.

8.5.3.3 Temporality

Diseases change and evolve in unpredictably surprising ways over time. Many modern neural network models, such as those developed recently in the medical profession, assume fixed quaternion data, making it impossible to control the time factor intuitively. A major component that will need the construction of new solutions is the development of deep learning algorithms that can truly manage temporal healthcare information.

8.5.3.4 Complexity of the Domain

Biomedicine and medical care involve more challenging topics than other application fields (such as picture and audio analysis). Because the bulk of diseases are so distinct, we do not know everything there is to know about their etiology and development. Furthermore, the number of patients in a real-world clinical setting is typically restricted, and we cannot always demand as many as we would need.

8.5.3.5 Interpretability

Despite the fact that deep learning models have worked effectively in a variety of applications, they are frequently considered black boxes. Although it may not be a concern in more dependable domains like picture annotation (since the final user may independently check the labels allocated to the pictures), in healthcare services, not just statistical computational performance but also the rationale why the algorithms function is vital. In practice, model exactness (i.e., determining which phenotypes drive projections) is crucial for convincing medical professionals to engage in the actions recommended by predictive systems.

8.6 Conclusion and Future Scope

In this chapter, an approach to the classification of skin cancer using base CNN, custom CNN and VGG16 along with a detailed discussion of how SDN, IoT implemented with deep learning makes healthcare as a whole much more efficient and smarter is presented. Exploration results show that techniques involving deep learning have the most noteworthy exactness and can be regarded as baseline methods with accuracy as high as 89%. The simulation was executed on Google Collaboratory, which has an Intel(R) Xeon(R) processor running at 2.30GHz, 13GB of RAM, and an NVIDIA Tesla K80 GPU. Further, various analyses have been presented on the effect of deep learning (its applications, challenges, etc.) on healthcare sector tweets. Use of computerized advances, for example, AI along with integration of IoT and SDN in the medical care field is entering an energizing period. The confluence of informatics, science, designing, science, and software engineering will quickly speed up our insight into both innate and ecological components adding to the beginning of complex infections. The capability of using duplicate number varieties in the expectation of malignant growth determination is energizing. Using AI to make an interpretable technique for seeing how the genomic scene interlinks across qualities to add to acquired malignant growth hazard might actually improve patient medical care on an individual level. In the near future, we intend to focus on the combination of machine learning and deep learning methods along with IoT-SDN in order to improve the general performance and overall condition of the medical sector.

References

1. van Kranenburg, R., and Dodson, S. (2008). *The Internet of Things: A Critique of Ambient Technology and the All-Seeing Network of RFID*. Institute of Network Cultures, Amsterdam University.
2. Da Xu, L., He, W., and Li, S. (2014). Internet of things in industries: A survey. *IEEE Transactions on Industrial Informatics*, 10, 2233–2243.
3. Kirichek, R., Vladyko, A., Zakharov, M., and Koucheryavy, A. (2016). Model networks for internet of things and SDN. In *2016 18th International Conference on Advanced Communication Technology (ICACT)*.
4. Kashani, M.H., Madanipour, M., Nikravan, M., Asghari, P., and Mahdipour, E. (2021). A systematic review of IoT in healthcare: Applications, techniques, and trends. *Journal of Network and Computer Applications*, 103164, Volume 192.
5. Sahoo, K.S., and Puthal, D. (2020). SDN-assisted DDoS defense framework for the Internet of multimedia things. *ACM Transactions on Multimedia Computing, Communications, and Applications (TOMM)*, 16(3s), 1–18.
6. Lecun, Y., Bottou, L., Bengio, Y., and Haffner, P. (1998). Gradient-based learning applied to document recognition. *Proceedings of the IEEE*, 86, 2278–2324.

7. Williams, R.J., Zipser, D. (1989). A learning algorithm for continually running fully recurrent neural networks. *Neural Computation*, 1, 270–280.

8. Smolensky, P. (1986). Information processing in dynamical systems: Foundations of harmony theory (No. CU-CS-321-86). Colorado University at Boulder Department of Computer Science.

9. Hinton, G.E., Salakhutdinov, R.R. (2006). Reducing the dimensionality of data with neural networks. *Science*, 313, 504–507.

10. Lieu, S., Liu, S., Cai, W., Pujol, S., Kikinis, R., and Feng, D. (2014). Early diagnosis of Alzheimer's disease with deep learning. In *International Symposium on Biomedical Imaging* (pp. 1015–1018). Beijing, China.

11. Esteva, A., Kuprel, B., Novoa, R.A., Ko, J., Swetter, S.M., Blau, H.M., and Thrun, S. (2017). Dermatologist-level classification of skin cancer with deep neural networks. *Nature*, 542, 115–118.

12. Toh, C., and Brody, J.P. (January 14th 2021). *Applications of Machine Learning in health-care, Smart Manufacturing – When Artificial Intelligence Meets the Internet of Things, Tan Yen Kheng.* IntechOpen. DOI: 10.5772/intechopen.92297.

13. Dernoncourt, F., Lee, J.Y., Uzuner, O., Szolovits, P. (2017). De-identification of patient notes with recurrent neural networks. *Journal of the American Medical Informatics Association*, 24(3), 596–606.

14. Zhou, J., and Troyanskaya, O.G. (2015). Predicting effects of noncoding variants with deep learning-based sequence model. *Nature Methods*, 12, 931–934.

15. Kelley, D.R., Snoek, J., Rinn, J.L. (2016). Basset: learning the regulatory code of the accessible genome with deep convolutional neural networks. *Genome Research*, 26, 990–999.

16. Prasoon, A., Petersen, K., Igel, C., Lauze, F., Dam, E., and Nielsen, M. 2013. Deep feature learning for knee cartilage segmentation using a triplanar convolutional neural network. In *International Conference on Medical Image Computing and Computer-Assisted Intervention* (Vol. 16, pp. 246–253).

17. Liu, S., Wang, Y., Yang, X., Lei, B., Liu, L., Li, S.X., Ni, D., Wang, T. (2019). Deep learning in medical ultrasound analysis: A review. *Engineering*, 5(2), 261–275.

18. Nguyen, P., Tran, T., Wickramasinghe, N., Venkatesh, S. (2017). Deepr: A convolutional net for medical records. *IEEE Journal of Biomedical and Health Informatics*, 21, 22–30.

19. Choi, E., Bahadori, M.T., Schuetz, A., Stewart, W.F., and Sun, J. (2016). Doctor AI: Predicting clinical events via recurrent neural networks. ar Xiv 2015.

20. Koh, P.W., Pierson, E., and Kundaje, A. (2017). Denoising genome-wide histone ChIP-seq with convolutional neural networks. bioRxiv 2016.

21. Kumar, A., Krishnamurthi, R., Nayyar, A., Sharma, K., Grover, V., and Hossain, E. (2020a). A novel smart healthcare design, simulation, and implementation using healthcare 4.0 processes. *IEEE Access*, 8, 118433–118471.

22. Pramanik, P.K.D., Pareek, G., and Nayyar, A. (2019). Security and Privacy in Remote Healthcare: Issues, Solutions, and Standards. In *Telemedicine Technologies*, Editors: Hemanth D. Jude, Valentina Emilia Balas, (pp. 201–225). Academic Press.

23. Pramanik, P.K.D., Solanki, A., Debnath, A., Nayyar, A., El-Sappagh, S., and Kwak, K.S. (2020). Advancing modern healthcare with nanotechnology, nanobiosensors, and internet of nano things: Taxonomies, applications, architecture, and challenges. *IEEE Access*, 8, 65230–65266.

24. Ali, S., Hafeez, Y., Jhanjhi, N.Z., Humayun, M., Imran, M., Nayyar, A., Singh, S., and Ra, I.H. (2020). Towards pattern-based change verification framework for cloud-enabled healthcare component-based. *IEEE Access*, 8, 148007–148020.

25. Mahapatra, B., Krishnamurthi, R., and Nayyar, A. (2019). Healthcare Models and Algorithms for Privacy and Security in Healthcare Records. In *Security and Privacy of Electronic Healthcare Records: Concepts, Paradigms and Solutions* (p. 183). IET. England & Wales.

26. Nayyar, A., Gadhavi, L., and Zaman, N. (2021). Machine Learning in Healthcare: Review, Opportunities and Challenges, *Machine Learning and the Internet of Medical Things in Healthcare*, Editors: Krishna Singh, Mohamed Elhoseny, Akansha Singh, Ahmed Elngar, Academic Press, (pp. 23–45). ISBN 9780128212295.

27. Krishnamurthi, R., Gopinathan, D., and Nayyar, A. (2021). A Comprehensive Overview of Fog Data Processing and Analytics for Healthcare 4.0, *Fog Computing for Healthcare 4.0 Environments*. 10.1007/978-3-030-46197-3_5. (pp. 103–129).

28. Nayyar, A., Mahapatra, B., Le, D., and Suseendran, G. (2018). Virtual Reality (VR) & Augmented Reality (AR) technologies for tourism and hospitality industry. *International Journal of Engineering & Technology*, 7(2.21), 156–160.

29. Azees, M., Vijayakumar, P., Karuppiah, M., and Nayyar, A. (2021). An efficient anonymous authentication and confidentiality preservation schemes for secure communications in wireless body area networks. *Wireless Networks*, 27(3), 2119–2130.

30. Alzubi, J., Nayyar, A., and Kumar, A. (November 2018). Machine learning from theory to algorithms: an overview. In *Journal of Physics: Conference Series* (Vol. 1142, No. 1, p. 012012). IOP Publishing.

31. Kumar, A., Sangwan, S.R., and Nayyar, A. (2020b). Multimedia Social Big Data: Mining. In *Multimedia big data computing for IoT Applications*, Editors: Sudeep Tanwar, Sudhanshu Tyagi, Neeraj Kumar, (pp. 289–321). Springer, Singapore.

32. Hakiri, A., Berthou, P., Gokhale, A., and Abdellatif, S. (2015). Publish/subscribeenabled software defined networking for efficient and scalable IoT communications. *IEEE Communications Magazine*, 53, 48–54.

33. Qin, Z., Denker, G., Giannelli, C., Bellavista, P., and Venkatasubramanian, N. (2014). A software defined networking architecture for the internet-of-things. In *Network Operations and Management Symposium (NOMS)* (pp. 1–9). IEEE.

34. Hu, F., Hao, Q., and Bao, K. (2014). A survey on software-defined network and OpenFlow: From concept to implementation. *IEEE Communication Surveys and Tutorials*, 16, 2181–2206.

35. Brewka, G. (1996). Artificial intelligence—A modern approach by Stuart Russell and Peter Norvig, Prentice Hall. Series in Artificial Intelligence, Englewood Cliffs, NJ. *The Knowledge Engineering Review*, 11(1), 78–79.

36. Kotsiantis, S.B. (2007). Supervised machine learning: A review of classification techniques. *Informatica*, 31(3), 249–268.

37. Gubbi, J., Buyya, R., Marusic, S., and Palaniswami, M. (2013). Internet of Things (IoT): A vision, architectural elements, and future directions. *Future Generation Computer Systems*, 29, 1645–1660.

38. Kim, H., and Feamster, N. (2013). Improving network management with software defined networking. *IEEE Communications Magazine*, 51, 114–119.

39. Libbrecht, M.W., and Noble, W.S. (2015). Machine learning applications in genetics and genomics. *Nature Reviews Genetics*, 16(6), 321–332.

40. Alpaydin, E. (2014). *Introduction to Machine Learning* (Vol. 3, p. 640). The MIT Press, London.

41. Längkvist, M., Karlsson, L., and Loutfi, A. (2014). A review of unsupervised feature learning and deep learning for time-series modeling. *Pattern Recognition Letters*, 42(1), 11–24.

42. Hazan, E., Klivans, A., and Yuan, Y. (2018). Hyperparameter optimization: A spectral approach. In *6th International Conference on Learning Representations, ICLR 2018; Conference Track Proceedings*.

43. Bardenet, R., Brendel, M., Kégl, B., and Sebag, M. (2013). Collaborative hyperparameter tuning. In *30th International Conference on Machine Learning; ICML*.

44. Song, C., Lee, H., Kang, C., Lee, W., Kim, Y.B., and Cha, S.W. (2017). Traffic speed prediction under weekday using convolutional neural networks concepts. In *IEEE Intelligent Vehicles Symposium (IV)* (pp. 1293–1298).

45. Chan, T.-H., Jia, K., Gao, S., Lu, J., Zeng, Z., and Ma, Y. (2015). PCANet: A simple deep learning baseline for image classification?. *IEEE Transactions on Image Processing*, 24(12), 5017–5032.

46. Simonyan, K., and Zisserman, A. (2014). Very deep convolutional networks for large-scale image recognition. arXiv 1409.1556.

47. Neha, B., Panda, S.K., Sahu, P.K., Sahoo, K.S., and Gandomi, A.H. (2022). A systematic review on osmotic computing. *ACM Transactions on Internet of Things*, 3(2), 1–30.

48. Sahoo, K.S., Tiwary, M., Luhach, A.K., Nayyar, A., Choo, K.K.R., and Bilal, M. (2021). Demand-supply based economic model for resource provisioning in industrial IoT traffic. *IEEE Internet of Things Journal*, vol. 9, no. 13, pp. 10529–10538, 1 July, 2022, doi: 10.1109/JIOT.2021.3122255.

49. Bhoi, A., Nayak, R.P., Bhoi, S.K., Sethi, S., Panda, S.K., Sahoo, K.S., and Nayyar, A. (2021). IoT-IIRS: Internet of Things based intelligent-irrigation recommendation system using machine learning approach for efficient water usage. *PeerJ Computer Science, 7,* e578.
50. Tripathy, H.K., Mishra, S., Suman, S., Nayyar, A., and Sahoo, K.S. (2022). Smart COVID-shield: an IoT driven reliable and automated prototype model for COVID-19 symptoms tracking. *Computing 104,* 1233–1254 (2022). https://doi.org/10.1007/s00607-021-01039-0

9

Applications of Machine Learning Techniques in SD-IoT Traffic Management

Anchal, Pooja Mittal, Preeti Gulia and Balkishan
DCSA, India

CONTENTS

9.1 Overview

The field of internet of things (IoT) communication technologies has seen tremendous expansion recently. Among the problems addressed by IoT development and research are quality of service requirements, massive data arrival, heterogeneous communication, and

DOI: 10.1201/9781003213871-9

unexpected network circumstances. Software-defined network applications are a significant contribution to the research world because they aim to organize rule-based management to regulate and add intelligence to the network by utilizing high-level policies that act as the network's controller by concealing low-level configuration issues. When machine learning techniques are coupled with software-defined networking, it is possible to create networking decisions that are more powerful and intelligent. IoT applications are now implementing resource virtualization and network control using SDN, which will assist traffic to become more regulated and sustainable. However, for the change to be viable, the IoT and the SDN requirements must be aligned. The main focus of this chapter is to discuss the uses of software-defined networking that makes use of IoT-enabled services to aid traffic control. In addition, the challenges and concerns by using deep learning and software capabilities in IoT applications is also discussed.

9.1.1 Introduction

The IoT reflects the internet's present and future status. The enormous quantity of connected things or objects generates a vast volume of data that requires multiple work and handling procedures for conversion into meaningful data. Moreover, to speed up and enhance the performance of the IoT network, advanced concepts in architecture and network administration are required to manage and regulate this huge quantity of data. Software-defined systems (SDSys) are a relatively new concept that aims to hide the difficulty with the architecture of conventional systems by placing the underpinning hardware in the middleware or software layer [1].

Most, if not all, individuals nowadays use the internet to fulfill personal, employment and commercial requirements. To do so, they must interface with a variety of gadgets and things. This communication between objects and humans requires the items around us to be connected to the internet. According to Gubbi et al. [2], IoT paradigm represents the existing and future state of the globe. Indeed, IoT researchers have predicted that the expansion of the internet of things to include all items in our surroundings will result in what they refer to as the internet of everything (IoE). The rapid expansion of the IoT is creating a huge quantity of information and data, which is gathered by the vast number of connected items. Storing, regulating, and protecting such large amounts of data are all essential concerns for linking everything with the internet in a practical and meaningful way. Furthermore, in real-time transactions or any basic task, it is necessary to link these things with one another to complete the task. Any delay in the communication response will have a detrimental effect on the overall accuracy and performance of the system. Solutions to speed up the communication process are seen as essential for IoT adoption and growth. SDSys are seen as a critical answer to these problems, their main objective is to keep the complexity of system resource management and control hidden from end-users.

For the internet of things [3, 4], traffic management is key to optimizing performance. Forecasting, dynamic analysis, and alteration of transmission data have all been used to improve execution using traffic management technologies. Currently available methods for traffic management are based on closed, inflexible network architectural designs in which data plane and control plane are strongly integrated, making it difficult to deliver real differentiated services for the large-scale IoT to handle rapid, uneven, increasing, and variable high-speed flows. A new concept of SD-IoT incorporates a software distinct structure into the architecture of IoT [5, 6]. Users may receive a dynamic, automated network with the software-defined framework, which is distinct from the past and enables complete virtualization application.

SD-IoT architecture splits the data plane from the control plane, which has apparent features such as visibility, virtualization, openness, programmability, and novel flow models and characteristics. Additionally, SD-IoT delivers a single international network perspective that opens the way for large-scale IoT traffic control and management solutions that are intrinsically flexible, adaptable, and configurable. SD-IoT architecture is divided into two parts: SD-IoT controllers and switches. The initial packet of a new flow is transmitted to the matching SD-IoT controller when it is received by an SD-IoT switch. To calculate the forward path for the flow, the SD-IoT controller is used. A new forwarding rule is installed on all SD-IoT switches on this path. The SD-IoT switch sends data to the SD-IoT controller, causing delays in transmission. The SD-IoT controller regulates the flow's forward route, causing processing delays. Even more concerning is the fact that installing new forwarding rules on forward-path switches takes a long time and is prone to delay peaks.

When a huge number of innovative flows are introduced into the SD-IoT switch, the data and control planes of the SD-IoT architecture both incur significant processing and communication costs. Beacon [7], the most powerful multi-threaded controller, can handle up to 12.8 Mfrps. By the end of 2016, worldwide M2M traffic had reached 32407.4 bbps. A separate SD-IoT controller has proved ineffective in meeting the needs of large-scale networks' massive data flow. The research in [8, 9] looked at data plane load balancing in large-scale SDNs to address the aforementioned issues of massive-scale IoT.

The overall objective of the study is to discuss various applications of machine learning techniques to manage SD-IoT infrastructure and the future potential to generate enhanced management for SD-IoT-based networks.

9.1.2 DDoS Detection

Web service attacks like DDoS commonly occur because they are easy to do and very effective. A DDoS attack, for example, has the power to cut off a country from the web. It is the goal of a DDoS attack to stop legitimate users from accessing services by using up all of the hardware resources or bandwidth, so that servers are unavailable. Machine learning is a good tool for cybersecurity because it can help users make the right decision and even do the right thing automatically. It includes artificial neural networks, fuzzy association rules, the Bayesian network, the clustering of data, decision trees, and evolutionary computation [10].

9.1.3 Software-Defined Networking (SDN)

Software-defined networking has assisted in the resolution of some difficulties in the networking sector. It decouples the control and data forwarding planes. As a result, SDN is more powerful than conventional networking. However, energy costs add to the total cost of the network. The router serves as both the control and data planes in a typical network. The router refreshes the routing table and obtains network status information in the control plane. In the data plane, the device routes incoming packets according to the routing table information. In SDN, the controller determines the routing and forwarding paths for all incoming packets, allowing the controller to also determine the usage of each node and connection. The switch lacks decision-making skills; it only operates as a forwarding device, passing packets according to the controller's instructions. Each switch is either active or inactive, as determined by the controller depending on the condition of incoming traffic [11].

9.1.4 Edge Computing

Edge computing refers to techniques that empower platform and IoT programs and services to still process at the program's edge, both on transmit and receive data. The IoT-based edge computing system architecture is classified into four divisions, as indicated in Figure 9.1: network domain, data context, software requirements, and device directory.

Sensors, meters, robots, and machine tools are examples of field devices that are anchored either in or near the device domain. The device domain must be able to create standardized transmission models to support a variety of transmission protocols and increase the flexibility of the transmission infrastructure. The device domain allows for real-time connectivity and smart deployment of field equipment and serves as a crucial basis for the application's upper layers [12].

9.1.5 Fog Computing

Edge computing has a limited quantity of resources, resulting in increased process delay and resource contention. This has led to the emergence of a new platform called fog computing, in which edge devices are integrated with cloud services in order to overcome the limitations of edge computing [13]. Fog computing is the use of computer resources for preliminary data processing and local storage in order to reduce network congestion or load. It provides localized data processing for quick decision making.

9.1.6 A Generic Framework for SD-IoT

As shown in Figure 9.2, basic devices are responsible for device resource management in an IoT environment [14].

IoT devices in the same region may communicate with one another through the region's basic controller. For one of the base controllers, each switch will act as a master controller, while the others will act as slave controllers. Through a central controller, IoT devices in various locations interact with one another. The basic controller connects with the data forwarding layer's switch, sending packet-out messages regularly, and receiving switch information all by feedback packet in messages. The basic controller interacts with the main controller to submit its control information so that the main controller gains a global picture of the whole network. Dynamic load balancing at the layer of the base controller and the coordination of messages across primary controllers are used to accomplish network load balance.

9.1.7 Traffic Management System Based on Software-Defined IoT

In big and developing metropolitan centers, traffic congestion is unavoidable, causing a slew of issues for individuals and groups. Emergency vehicle movement is impeded by traffic congestion, which costs lives. As time has passed, the network has evolved significantly from a traditional IP system to the present developing networking model known as a software-defined network. Before we go into the new networking concept, let us look at some of the old network's shortcomings [15] as follows:

- Traditional IP networks are difficult to operate and maintain.
- Using vendor-dependent commands, the system administrator must set up each device independently.

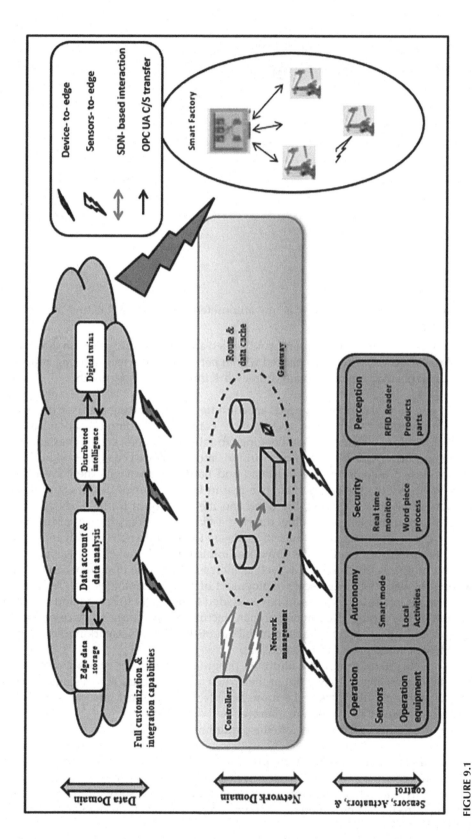

FIGURE 9.1
IoT-based manufacturing architecture of an edge computing platform.

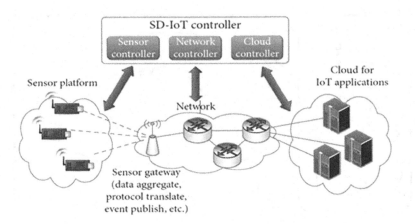

FIGURE 9.2
A generic framework for SD-IoT.

- In today's IP networks, there is no automated reconfiguration and reaction mechanism.
- The control plane and information plane (which advances traffic based on the control plane's decisions) are both packed within networking devices, limiting adaptability, and obstructing networking framework innovation and growth.

The quantity and variety of network needs are continuously increasing. Because of the network's dynamic and static structure, adding more controls and communication capacity with different vendor equipment is challenging. To satisfy demands such as variable QoS levels, large and unpredictable volumes of traffic, and security needs, networking infrastructure has grown increasingly complicated and harder to manage. The conventional internet is dealing with challenges such as a rise in the number of devices, an increase in the amount of cloud computing, big data, and the velocity of traffic. SDN can make the internet programmable and flexible, offering a viable solution to many of the issues that plague today's internet architecture. Its goals are to integrate network advances more quickly and to simplify and automate network administration. It makes testing and implementing new protocols much easier, as well as increasing the rate of network innovation and lowering the barrier to large-scale deployment of new technologies [16]. As a result, big firms such as Google and Microsoft have decided to use SDN for cloud and data center operations. It has piqued the interest of manufacturers and internet service providers. Nokia, Cisco, and Ericsson are all active members of the Open Networking Foundation (ONF) for SDN development and research [15].

SDN is a novel system design that improves programmability and centralized network intellect by taking the control plane out of the switch and delegating it to a single piece of apparatus, allowing the network to be more flexible. Over the last few years, SDN has aimed to change the way networks are managed and designed.

The following points will demonstrate the requirements of SDN:

- **Virtualization**: While network virtualization enables organizations to create multiple virtual networks within a single physical network or to connect devices from different physical networks to form a single virtual network, software-defined networking introduces a new method of controlling data packet routing via a centralized server.

- **Orchestration**: Network orchestration is the act of automatically configuring the network's behavior in such a way that it works in unison with hardware and software to enable additional applications and services.
- **Programmable**: A programmable network is one in which the behavior of network devices and flow control is managed by software that is completely separate from the network hardware. A genuinely programmable network enables network engineers to re-program existing network infrastructure rather than rebuilding it manually.
- **Visibility**: SDNs enhance network visibility in multi-domain locations thanks to their ability to access all the elements of a network through the visibility plane.
- **Dynamic Scaling**: Dynamic scaling is a litmus test that shows whether an evolving system exhibits self-similarity. In general, a function is said to exhibit dynamic scaling if it satisfies [15].

Organization of Chapter

The chapter is organized as: Section 9.2 and 9.3 highlights the generic concepts of IoT and IoT-based Traffic Management. Section 9.4 enlightens machine learning and Smart Transportation is focused at section 9.5. Section 9.6 concludes the chapter.

9.2 IoT

IoT has a distinct identity helping it to communicate and interact with other things. IP networks, which employ several communication protocols, are used to connect IoT devices. Data delivery is handled via the cloud-based intermediate layer. In this case, the devices require an active internet connection. It allows for an endless number of integration options, but it requires a way of handling all communication. Collaborative driving is a concept that enhances road transport safety while also reducing the number of fatalities and injuries in car accidents. This accomplishment is based on data exchanged between automobiles equipped with sensors and other devices that allow them to sense their surroundings and participate in dynamically formed groups. Localized networks are a collection of cars that may design a collective driving strategy that requires little or no driver input. Most architectures have not addressed, or have only partially addressed, the issue of inter-vehicle communication [17].

Figure 9.3 shows the different layers of IoT architecture: network, perception and application. IoT requires a smart connection with context-aware and existing computing networks leveraging network resources. The development of ubiquitous communication and information networks is already visible, with the rising availability of Wi-Fi and 4G-LTE wireless internet access. However, in order for the IoT goal to succeed, computing criteria need to expand past typical mobile computing scenarios including portables and smartphones, to include linking common existing things and integrating intelligence into our surroundings. The IoT necessitates the disappearance of technology from the user's mind. Smart connection and context-aware computing may be achieved with these three basic foundations in place:

(1) A piece of common knowledge about the situations of users and their equipment;
(2) Software architecture and ubiquitous communication networks utilized to process and send contextual data when needed;
(3) IoT analytics tools that strive for self-driving and intelligent behavior.

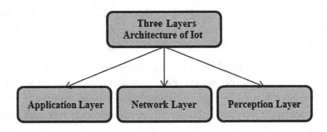

FIGURE 9.3
Three-layer IoT architecture.

FIGURE 9.4
Five-layer architecture of IoT.

The present web is being converted into a network of linked devices that not only gather and interact with the environment (sensing) data but also utilize existing internet standards to offer data transmission, analysis, apps, and communications services. The ubiquity of gadgets that can be accessed via open wireless technologies like Bluetooth, RFID, Wi-Fi and telephone data, embedded sensors, and actuator nodes have driven IoT beyond its infancy and are about to convert the present static internet into a fully integrated 'future internet' [18]. Individuals now have unprecedented levels of connection and speed because of the internet revolution. The next revolution will be the integration of things to create an intelligent environment.

When a task necessitates the utilization of a variety of cutting-edge techniques and a wide range of application areas, a five-layer design is the optimal option. The five-layer concept may be seen as an extension of the IoT architecture (see Figure 9.4) [19].

- **Perception Layer**: This is the first layer of the IoT architecture. It collects essential data such as moisture content, temperature, vibrations, and intrusion detection using a variety of sensors and actuators. The main purpose of this layer is to acquire data from the environment and transfer it to another layer so that activities can be performed in response to the data.
- **Network Layer**: As the name indicates, this is a layer that links the perception and middleware layers. Using network technology including 3G, 4G, UTMS, infrared, and Wi-Fi, it collects data from the perception layer and sends it to the middleware layer. Even though it enables the communication between middleware and awareness levels, the communication layer is also known as the interface layer.
- **Middleware Layer**: The middleware layer adds additional capabilities such as computing, storage, processing, and decision making. It stores all data and transfers it to the appropriate computer depending on the location and identity of the device. It may also reach conclusions based on the results of calculations conducted on sensor data sets.
- **Application Layer**: The application layer handles all application operations, such as setting alarms, sending emails, turning on and off devices, security systems,

smart agriculture, and wearables, using the information received from the middle-ware layer.

- **Business Layer**: Among other things, it generates flowcharts and diagrams, evaluates data, and decides how the gadget may be improved [19].

Furthermore, IoT devices perform basic tasks such as data collection, communication with M2M, and preprocessing of data, according to apps. A balance between costs, processing capabilities, and energy use is important while designing or choosing an IoT device. Since IoT devices gather and exchange a huge amount of data continuously, IoT is closely related to big data. As a consequence, an IoT infrastructure typically includes methods to handle, store, and analyze massive quantities of data. [20].

IoT systems have become widely distributed practices [21]. IoT systems also include features like monitoring, node administration, data storage and analysis, data-driven customizable policies, and more. Depending on the application, it may be necessary to do certain data processing tasks in IoT devices rather than a centralized node as in the cloud computing infrastructure. When computing advances closer, a novel computer paradigm known as edge computing emerges [22].

Since these instruments are often low end, they may not be capable of handling intensive processing activities. Consequently, an intermediate node with satisfactory capacity and the ability to execute advanced processing tasks is required, and it must be physically situated near the end network modules to decrease the overload produced by the large transmission of all data. The introduction of fog nodes [23] provided the answer. Fog nodes provide storage, processing, and networking capabilities to IoT devices, assisting them in handling large amounts of data. The data is eventually saved in cloud servers, where it can be analyzed using several machine learning algorithms and shared with other devices, resulting in the development of new value-added smart apps. In many elements of the so-called smart city, IoT apps have already been developed. Among the most essential applications are the following (Figure 9.5) [24]:

Smart homes: Traditional home appliances, such as refrigerators, washing machines, and light bulbs, may now link with one other, or with authorized users on the internet, for improved supervision and administration of devices and optimization of energy consumption. In addition to conventional devices, new technologies such as intelligent home supports, intelligent port locks, etc. have been developed. [24].

Health-care assistance: Innovative gadgets have been created to improve patient well-being. Wireless sensor plasters can monitor the condition of an injury and communicate the information to a doctor. Other sensors, like wearable devices, may be used to monitor and report various data, like blood sugar level, blood oxygen, cardiac rate, or temperature [24].

Smart Transportation: With car-embedded sensors or mobile devices placed in the town, optimal route recommendations, simple parking facilities, economical street lighting, public transit telematics, accident avoidance, and self-contained driving may be offered. [24].

Environmental Conditions: Dispersed wireless sensors are ideal for monitoring a broad range of ambient variables. The creation of sophisticated weather stations may assist with barometers or ultrasound wind sensors. In addition, intelligent sensors can monitor water and air quality in the city [24].

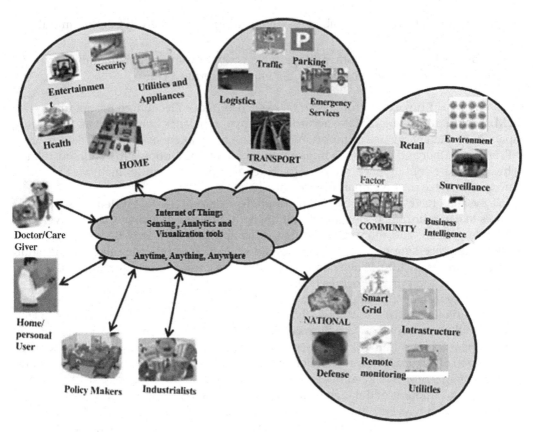

FIGURE 9.5
IoT showing the end-user and application areas based on data.

Logistics and Supply-Chain Management: Intelligent RFID tags make it easy to monitor goods from manufacture to shop, substantially saving cost and time. Intelligent packaging can provide brand quality, protection assurance, and customer customization. [24].

Security and Surveillance Systems: Intelligent cameras can gather images at various places on the street. Smart safety systems can identify crime via real-time visual object identification or prevent hazardous circumstances. There is still a great deal to be done with regard to standardization in IoT architecture and technology [24].

9.2.1 Cybersecurity and Internet of Things (IoT)

The internet of things simplifies our lives while risking our security. The fast growth of intelligent IoT devices has created several avenues for hackers to infiltrate user privacy. During data transfer, data may be accessed or altered by intruders or hackers, or by a variety of other factors such as a server breakdown or an electromagnetic disruption. Data integrity guarantees that intruders, hackers, or cybercriminals do not tamper with data during transmission. There is a lot of risk for commercial products in the cyber world. No security measures are completely up to date, and even if these existed, there is no way to avoid risk. Cybersecurity policy should be based on a good understanding of the cybersecurity threat, and the private sector will play just as important a role as the public sector [25].

9.2.2 IoT Terms and Acronyms

Machine-to-machine communication (M2M): M2M is a communiqué network between devices that use any channel. Initially employed in an industrial setting, communication was used for personal equipment transmission. The IoT is also communication among nodes, often referred to as the vertical program, and achieves communication between many gadgets. Table 9.1. underlines the key differences between M2M and IoT.

Cyber-physical systems (CPS): Cyber-physical systems contain interacting digital, analog, physical, and human components built for operation through integrated physics and logic, according to the NIST. These technologies underpin the essential infrastructure, serve as the platform for existing and intelligent skills, and enhance our quality of life in a variety of ways. CPS are used as part of many industrial processes. Beyond manufacturing, cyber-physical systems may be found in smart grids and smart cities, for example [26].

CISCO used the term 'Internet of Everything (IoE)' to describe people, process, data, and objects that make networked connections more relevant and useful, turning information into actions that enhance everything. In 2017, this phrase was phased out.

Social Internet of Things (SIoT): SIoT is an internet of things that may generate a system of social interactions without the need for human involvement. Objects can start forming social interactions based on their profile, interests (i.e., the apps they use, the services they utilize), and behaviors, i.e., movements. These social ties may also be established in order to develop events. For example, a collaborative link may be created between devices working together to generate mutual IoT apps, such as telemedicine or emergency response items. Because they are part of the same batch, items that have the same maker, are of the same model, or were built during the same time period may have a parental connection. Because the object owners are in contact, social ties are formed between items that come into contact sporadically or continually, and a co-ownership relationship may form among diverse objects owned by the same user. Implementation of the SIoT concept has several benefits [27]:

- The social network formed by SIoT objects may be designed to provide network browsability, effective item or service discovery, and scalability, such as human social networks.
- Confidence may be established to balance the amount of contact between friendly things.
- Social-networking models may be used to deal with IoT problems in linked item networks.

TABLE 9.1

Key Differences between IoT and M2M

IoT	M2M
Devices communicate using IP networks, varying communications protocols are possible	Point-to-point communications – embedded in hardware at the customer site
Data delivery is relayed through a middle layer in the cloud	Several devices use these protocols, over cellular networks
Active internet connection mandatory	No internet connection required
Integration options are more varied, but management is necessary	Limited integration options: devices must have communications standards

9.2.3 Features of IoT

The IoT is a network of connected devices. These items might be anything that consumers view and utilize regularly. They may also be clothing, trash cans, chairs, and other items that consumers would not normally consider electronic [28]. These are interconnected smart objects that can exchange resources and information. The information gleaned from the sources is gathered and investigated further [29]. Smart objects have computing power, sensors, and the ability to connect with other items. Smart objects are used to make ordinary tasks or labor easier for humans, and the IoT has provided various benefits to enhance people's quality of life [30]. The following are the major characteristics of the IoT:

- *Inter-connectivity and services*: IoT devices can be associated with the global ICT infrastructure, according to [31]. Under object constraints, IoT delivers characteristics such as privacy, semantic coherence, and protection between real items and their virtual interconnected selves. Technology is emerging to offer services related to objects in both the digital and physical worlds.

- *Heterogeneity*: The nature of IoT devices is diverse. This functionality enables several devices to communicate with each other across a variety of hardware platforms, networks, and technologies. The IoT system includes the use of different protocols to link a variety of devices, platforms, and operating systems [32]. The IoT links devices with different power requirements, sources, and purposes. Because the IoT's goal is to provide seamless M2M, M2H, and H2H connectivity, it must be able to connect a variety of devices and networks [33].

- *Dynamic changes*: Dynamic IoT devices, which means that they change constantly throughout time. For instance, the devices may be in a logged on/off or sleep/wake state. The number of gadgets to be used simultaneously varies. Certain devices may be added to the network, while others can be removed. IoT devices may adjust to different requirements. For example, security cameras may change their settings during the daytime [32]. The connection of the device with other devices may vary over time so that one set of devices may be linked to another, the next is the need for efficient security procedures and cryptographic systems to guarantee security [33].

- *Enormous Scale*: The number of networked gadgets to be monitored and managed would be at least more than the number of presently linked internet gadgets. It will be far more difficult to use and understand the large quantities of information produced by these devices in various applications. [31].

- *Safety*: While the IoT has many advantages, it also poses serious safety concerns. Whether they are IoT receivers or producers, users must ensure their safety. It includes both physical well-being and personal information about the individual. It entails safeguarding the security of communiqué endpoints, networks, and data in transit, resulting in a scalable security standard [35].

- *Connectivity*: The most essential factor to take into consideration in the context of IoT is connectivity. No business use case is possible without smooth communication among the interconnected elements of IoT ecosystems (such as sensors, compute engines, data hubs, etc.). IoT devices can connect via radio waves, Bluetooth, wi-fi, and other technologies. To enhance efficiency and promote universal communication across IoT ecosystems and industry, we can utilize several internet connectivity layer protocols.

- *Green Internet of Things (G-IoT):* The phrase 'green IoT' refers to energy-efficient operations (hardware or software) done via IoT to allow for the reduction of the greenhouse impact generated by present apps and services or to reduce the IoT's greenhouse effect. To have low or no environmental impact, the whole lifecycle of green IoT should be centered on green design, green production, green usage, and finally green disposal/recycling [34].

9.3 IoT-Based Traffic Management

Population growth is posing a number of worrying difficulties for the world's metropolitan regions. Waste management, resource scarcity, air pollution, human health concerns, and transportation congestion are all issues that arise as a result of such massive and diverse crowds. The concept of smart cities was developed to address challenges associated with the rising urban population. Smart cities are defined in a variety of ways: the use of smart computing technology to make the key infrastructure components and services of a city, according to one definition. The IoT is a recently emerged archetype that seeks to give new communication and IT options. Virtually everything will be connected to the internet under the IoT concept. As a result, the IoT can play a critical role in smart cities. To connect devices, IoT requires cloud computing infrastructure. IoT infrastructure may be utilized for embedded PCs such as Galileo, Raspberry Pi Foundation and Intel, or similar embedded computer units [36–38, 40].

This section suggests an IoT-based traffic control organization for smart towns. In any big city, traffic congestion is a serious issue. According to research, people in the United States spend around 38 hours per year stuck in traffic [39]. Modern traffic signals were invented in the United States, and a traffic signaling manual outlines contemporary traffic signaling standards. The manual control for three-color traffic signaling systems originally came from a tower in the middle of the roadway. Figure 9.6 depicts a typical contemporary traffic light model.

The traffic light control arrangement was originally designed around a fixed signaling method [41]. The concept of making traffic signal management adaptable based on real-time traffic volume was first proposed in 1982, and it has since been executed in several major cities. The adaptive traffic signal system outperformed first-generation traffic control in terms of traffic congestion management [42]. Fully adaptive and dynamic decision making is a feature of the third generation of traffic signaling control. The traffic signaling systems are adjusted in real time to reflect the current traffic conditions at an intersection. Most traffic light control systems use microprocessors and follow a preprogrammed algorithm. Changing the traffic light control patterns depending on the traffic circumstances is generally required [43].

These techniques seek to provide automated traffic regulation by intelligently responding to specific situations. To solve the traffic congestion problem, an advanced prototype, where the number of vehicles and velocities are transmitted to a backend framework is proposed, which conjectures and provides options for setting red or green light duration via a traffic light control interface. Strategies to manage traffic congestion are outlined in a study released by the University of New Mexico. All these signaling methods are expensive to implement. Furthermore, environmental factors might have an impact on their functionality.

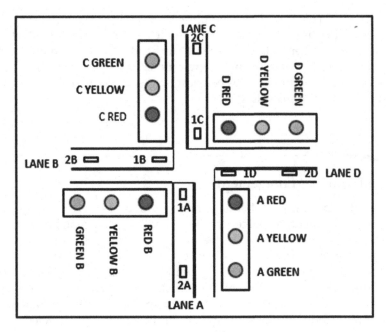

FIGURE 9.6
Traffic signal model.

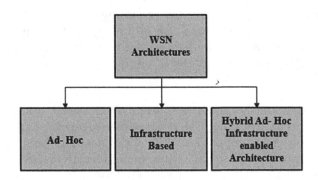

FIGURE 9.7
WSN architecture.

The next generation of smart traffic controllers, on the other hand, is still being evaluated and developed. Experts are concentrating on the concept of smart cities, which aims to handle all challenges that arise in large cities, including traffic congestion.

9.3.1 Growth of IoT Devices for Traffic Management System

Traffic congestion is a serious problem that affects towns all over the globe, with negative consequences including increased fuel use, air pollution, and slower economic growth

[45]. Malaysia's yearly economic losses due to traffic congestion are estimated to be over RM20 billion. [46]. The major causes of traffic congestion are population growth and a lack of public transportation. The number of urban residents in Malaysia has risen by 242% in the last 30 years, while the number of automobiles has climbed by 585% in the last 35 years. In Malaysia, public transportation development is largely focused on the Klang Valley region, which has a population of around 7.2 million people [47, 48]. Despite the government's best efforts to alleviate traffic congestion in the Klang Valley region, other major towns such as Johor Bahru, George Town, Kota Kinabalu, and Kuching demand improvements in public transport, with the bus being the sole alternative for locals.

According to [50], Malaysia has a total road network coverage of 144,403 km, including 116,169 km of paved roads and 28,234 km of unpaved roads. However, without additional enlargement, the old, paved road are unable to deal with the increased traffic volume. To better regulate large traffic flows on ancient, paved roads, traffic lights have been installed at each road intersection. Fixed-cycle traffic signals are extensively used in Malaysia, and they tend to cause traffic congestion, especially on roads with significant traffic volume. To alleviate network congestion, a dynamic cycle traffic light system (TLS) is necessary.

There are three kinds of sensor communication networks in ad-hoc architecture: on-road sensor networks, on-vehicle sensor networks, and hybrid on-vehicle sensor networks. On-road sensor networks require all sensors to be communicating with one another in a multi-ad-hoc manner without the use of infrastructure. For direct vehicle-to-vehicle (V2V) communication, the on-vehicle sensor network uses the vehicle's in-built sensors. Vehicular ad-hoc networks can be regarded as a subset of MANETs. However, in recent years, they have evolved into a distinct scientific topic. VANETs and MANETs have comparable characteristics such as multi-hop communications, changing topologies, mobility, and power transfer constraints [41, 49, 51]. The hybrid on-sensor on-vehicle networks incorporates both communication techniques.

One of the sensors serves as the master, deciding on the green light phase time (GLPT) for each route. Figure 9.9 depicts several ad-hoc architectures [52].

Ad-hoc architecture has three kinds of sensor nodes with its base station (BS) network: on-road sensor network with BS, on-road sensors network with the BS, and hybrid on-road sensor network with the BS. The infrastructure-based BS network utilizes cellular, Wi-Fi, or DSRC to communicate with these sensors and then determines whether GLPT is used for highway traffic. Several architectures are illustrated in Figure 9.9 [52].

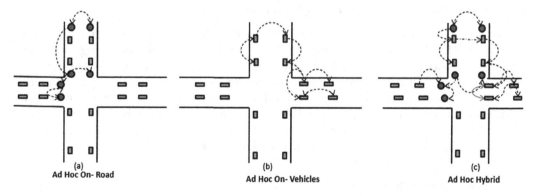

(a) Ad Hoc On- Road   (b) Ad Hoc On- Vehicles   (c) Ad Hoc Hybrid

FIGURE 9.8
Ad-hoc architecture.

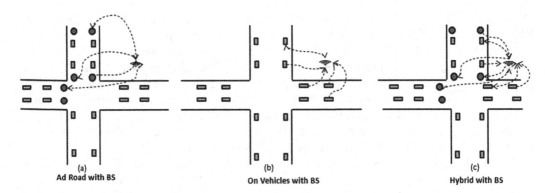

FIGURE 9.9
Infrastructure-based architecture.

9.3.2 Existing Traffic Signals

SCOOT and SCATS are the two most often utilized dynamically cyclic TLSs in the world [53]. Inductive loop detectors are used by both TLSs. These statistics are utilized to calculate the GLPT for each route using three traffic optimizers: offset, cycle, and split time. These optimizers may aid in reducing unnecessary green time at traffic crossings and in coordinating nearby sets of signals to reduce delays.

9.3.3 Algorithm for Calculating Green Light Phase Time

Various algorithms have been used by several researchers to optimize the GLPT. Yousef et al. [54] proposed that GLPT be calculated using the Traffic Signals Time Manipulation Algorithm (TSTMA) and Traffic System Communication Algorithm (TSCA). Collotta et al. [55] proposed a dynamic management algorithm. Next, gradient-based optimization using infinitesimal perturbation analysis (IPA) was proposed by Fleck et al. [56]. For GLPT computation, a fuzzy neural network approach was developed by Xiaohong et al. [57]. Aside from that, ensemble-based systems are proposed for GLPT computation [58]. For GLPT computation, a multi-objective particle swarm optimization method based on crowding detachment was suggested by Zhang et al. [59].

9.3.4 Wireless Traffic Controller during Rush Hour

At rush hour, traffic police will be assigned at junctions to manually control traffic flow by turning off TLS. This technique necessitates the police department allocating many officials during peak hours, putting their health at risk owing to air pollution. During busy hours, a wireless traffic light controller was suggested in [60], which can help traffic police regulate traffic lights. The controller, which is located next to the road intersection, has two modes: manual and automatic. Automatic mode changes the traffic light structure depending on pre-specified settings, whilst manual mode gives the traffic light settings required by the police, making it easier for them to maintain control.

9.3.5 Traffic Light System Security

Wireless communications allow all sensors and microprocessors at each traffic intersection to be connected, reducing hardware wiring expenses. Wireless communications, on the other hand, are frequently unprotected. A group of researchers in Michigan, United States, researched and identified security vulnerabilities in wireless TLS in [61]. The system has three main flaws: unencrypted wireless radio transmissions at 900 MHz and 5.8 GHz, network devices utilized factory-default passwords and usernames, and a controller debugging port that may be accessed at any time. Because of these flaws, an attacker can manipulate lights for personal benefit, causing traffic gridlock.

9.4 Machine Learning

Machine learning (ML) is a notion that has been around for a long time. AI and ML are closely connected. AI is made possible through ML. Computer systems can learn to execute responsibilities like pattern recognition, predictions, clustering, categorization, and more using ML. Systems are taught to examine sample data to archive the learning process utilizing numerous statistical and algorithm models. A machine learning algorithm seeks to identify a link among the topographies and certain output values called tags, and the sample data is generally defined by quantifiable qualities called features [62]. The data gathered throughout the teaching phase is then utilized to spot patterns or make choices centered on new data. Regression, classification, grouping, and determining association rules are all issues that ML excels at. ML algorithms may be divided into four types based on the learning style as shown in Figure 9.10:

i. *Supervised Learning*: Supervised learning uses algorithms like random forest and linear regression to solve issues requiring prediction, assessment of life experience, and population growth forecast. Additionally, supervised learning uses algorithms like SVM, nearest neighbor and random forest to solve classification issues including digit identification, audio recognition, diagnostics, and identity fraud detection. In supervised learning, there are two stages: training and testing. The data sets utilized in the training phase must have labels that are known. The algorithms aim to

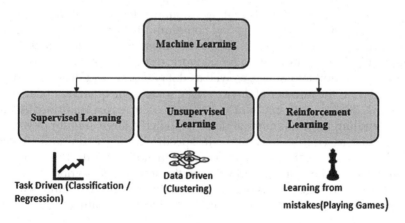

FIGURE 9.10
Types of machine learning.

anticipate the output values of the testing data by learning the connection between input values and labels. [63].

ii. *Unsupervised Learning*: Unsupervised learning is used to solve challenges like big data visualization, feature elicitation, and hidden structure detection. In addition, supervised learning is utilized to solve problems like recommend systems, consumer segmentation, and targeted marketing. In contrast to supervised learning, no labels are accessible. This group of algorithms aims to find patterns in testing data, cluster it, or forecast future values [63].

iii. *Semi-supervised Learning*: This is a hybrid of the two preceding techniques. Information that has been tagged and data that has not been labeled are both used. It functions in largely the same way as unsupervised learning, but with the added benefit of a small quantity of labeled data [62].

iv. *Reinforcement Learning*: The algorithms in this knowledge approach attempt to anticipate the result of the issue using a set of tuning parameters. The intended output is then used as an input parameter to generate fresh outputs until the ideal output is obtained. The learning techniques used are deep learning and ANN. Reinforcement learning is mostly utilized in applications such as artificial intelligence games, robot navigation, skill acquisition, and real-time decision making [63].

There are two key criteria to consider when utilizing ML techniques: how computationally intensive and how fast an approach is. The most appropriate ML algorithm is chosen based on the application type. If a real-time examination is required, for example, the algorithm used must be able to follow changes in the input data and generate the appropriate result in a precise way.

9.4.1 Machine Learning in IoT

IoT connects gadgets, allows them to communicate with one another, and gathers a massive amount of information. Human involvement is necessary to evaluate the acquired data, extract relevant data, and develop smart applications. IoT devices must be self-contained in addition to collecting data and communicating with other devices. They must be able to make judgments centered on the topic and learn from their data. The name cognitive IoT (CIoT) was coined to address this issue [64].

Intelligent IoT devices, capable of creating automated resource allocation by automated smart applications, network operation, and communication, are also required. The use of ML techniques in an IoT infrastructure can lead to substantial developments in infrastructure and applications. ML may be used for network, optimization, resource allocation optimization, and congestion avoidance, as well as data analysis and decision making in real time. Furthermore, as the number of devices rises, so does the quantity of data collected. In IoT applications, dealing with big data is a regular occurrence. Traditional databases are incapable of handling large amounts of data. To switch a large amount of structured and unstructured data, as well as to analyze it, specific infrastructure and procedures are required [65]. Many ML methods, like Ensemble, can aid in the effective handling of large quantities of data, and they will be addressed in the subsequent sections.

Classification: The arrangement is one of the most important ML methods in the IoT. ML classification techniques employ input-trained data to evaluate the likelihood or

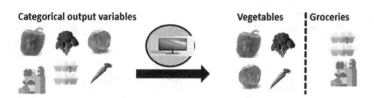

FIGURE 9.11
Classification of variables.

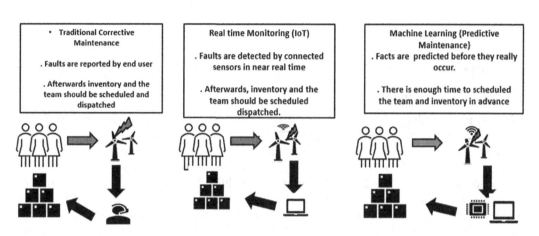

FIGURE 9.12
Predictive maintenance in the energy industry.

possibility that the data that follows will fall into one of the specified categories. Figure 9.11 depicts how the independent input variables are classified into categorical output variables.

Regression: Predictive maintenance (PdM) is a cutting-edge maintenance approach that has been used in a variety of industries. PdM's primary goal is to determine whether equipment requires supervision because components may fail in the future, as well as to forecast the residual usable life (RUL) of machine components (Figure 9.12). Regression methods may be used to correctly forecast RULs in these scenarios.

Clustering: Data processing at a plant manufacturing automobile engines is an example of unsupervised learning in IoT. Assume we want to create a machine that can identify engines that need to be adjusted further. Visually detecting faults is difficult, but it may be done by collecting many critical characteristics and applying a clustering procedure to identify groups. Constraints are generated by sound and temperature, for example, the clustering algorithm will classify the engines into various groups dependent on mutuality in generating alarm at a particular temperature limit (see Figure 9.13). This will assist the factory's engineers to rapidly identify the engines that belong to the issue group.

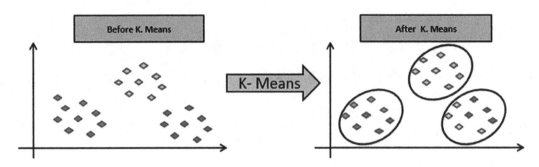

FIGURE 9.13
Clustering of engine data.

9.5 Smart Transportation

As IoT expertise advances, new requests to improve people's lives are developing. Smart city apps are being created to make use of the newest technology advances. Transportation systems begin to feel and think because of the advent of IoT in the area of transportation, leading to the creation of intelligent transportation systems (ITS). Many researchers are interested in smart transportation due to its possibilities for improvement. Navigation or route optimization is one of the extremely important topics of interest in smart transportation. The process attempts to predict traffic overcrowding and suggest optimum route choices to reduce driving times, and ultimately to decrease vehicle energy consumption and emissions, using mobile [66] or side units placed in specific places on the road [67] [68].

Streetlights that can identify traffic conditions and function appropriately, rather than being turned on and off on a schedule, are suggested to help reduce energy consumption. Smart parking systems have also benefited from the use of IoT devices. Researchers have suggested novel parking reservation systems that use cameras [69] such as infrared sensors [70] to optimize a parking lot's capacity and availability while reducing searching time. Furthermore, technologies that aid in the detection of road surface abnormalities based on data from vehicle sensors or the driver's phone have been suggested. There have also been attempts to use IoT devices to identify or prevent traffic accidents. Lastly, the IoT M2M communiqué way has enabled development networks in which cars may share valuable information and open a plethora of new application possibilities [71].

9.5.1 ML and IoT in Smart Transportation

Smart transportation in today's world has been widely studied due to the variety of daily issues and their significant impact on the contemporary smart city. Furthermore, the nature of the issues it addresses encourages the adoption of both ML and IoT technologies. As a result, the study focuses on the most current explored works that utilize IoT and/or ML methods to solve smart transportation classifications. The following sections classify ML procedures by algorithm type:

Ensemble: By training many classifiers and pooling their findings, ensemble algorithms solve a problem. The ability of ensemble techniques to enhance so-called weak classifiers to strong classifiers is their primary benefit. In this manner, systems may utilize weak classifiers that are simple to build, while achieving the quality of a strong classifier [72]. The creators of [73] want to avoid traffic accidents by developing a framework for detecting the driver's awareness. A camera mounted in front of the driver collects the input data by tracking head nodding and eye movement. Then, from an extracted integral picture, human HAAR characteristics [74] are chosen for the adaptive boosting algorithm. AdaBoost is an ensemble method that combines other weak classifiers to create a powerful classifier. Many characteristics in the study must be analyzed in real time, and AdaBoost has a significant benefit in this regard. Another prominent ensemble algorithm is RF. For classification and regression problems, RF employs a slew of decision trees as feeble learners, with the mean value or most tested values as the output. Single decision trees are more vulnerable to the over-fitting phenomenon than RF [75].

Bayesian: A directed acyclic graph (DAG) represents a Bayesian network and gears a chance supply intended for a collection of random variables. The model may be estimated using a training set as variables, and then used to allocate tickets to fresh data. A Bayesian network (BN) is a prototypical network that explains all the variables and how they relate to one another. As a result, seeing certain variables in a BN may provide data about the status of an additional set of variables [76].

Markov Models: Markov Models (MM) are unpredictable sequencing models based on probability distributions. A Markov chain is the most basic MM. Circulation of a variable that changes its value arbitrarily over time in a Markov chain is only dependent on the delivery of the preceding state. A hidden MM (HMM) is a Markov chain that utilizes hidden states. The hidden states, the number of yields, and the state transition probability dispersal may all be used to describe an HMM [77]. Liu et al. [78] dealt with motion intent interpretation by allowing cooperative vehicle observation. The perception range of the vehicle is extended thanks to cooperative sensing, resulting in improved perception quality and adequate reaction time for autonomously triggered actions. A simple form of coupled hidden MM (CHMM) is used to do this. The model is trained using velocity data from a vehicle sequence's different driving characteristics. A linked HMM is a kind of HMM that is better suited to situations involving several processes.

Decision Trees: Decision Trees (DT) are structures used to address organizational challenges that depict decision making with multiple options. A top-down DT is composed of multiple nodes that lead a testing object to a class. It is a crude solution to categorization issues. At each level of a decision DT, a new categorization sub-process occurs, dividing the primary job into smaller sub-tasks. As previously stated, a decision tree is used in [79] to forecast traffic congestion with four other ML methods. For precision, recall, and accuracy, the decision tree is on a par with the other methods, although logistic regression was somewhat superior. The authors used a regression tree (RT) to predict short- and long-term traffic flow in work zones, which was one of the ML methods used in [80]. RT is like decision trees, except that instead of a binary class, the tree's response variable is a numeric value. The squared regression error is then computed after a binary recursive partitioning. The branch containing the variables with the lowest sum of squared errors is selected.

Clustering: Clustering is an unsupervised technique for categorizing items into distinct patterns found. It has no recognized labels to train the archetypal in clustering, unlike classification techniques. To improve the traffic network design by calculating the optimum location and amount of processing centers for a given number of providers, [66] was used. To reduce costs, the authors utilize k-means to determine the distance between centers and suppliers. The real-time data obtained from GIS sensors is pre-processed using a deep belief network (DBN) method before k-means clustering. It is the DBN that discovers the first K locations for the k-means.

The suggested technique in [81] utilizes portable devices to gather data via speedometers to identify the position of road abnormalities such as speed bumps and potholes. After this, the data are transmitted to the closest fog node to be processed. Clustering the information and determining threshold values that differentiate a road anomaly from a typical road, k-means is employed. Ghadge et al. [82] proposed a similar method for detecting road abnormalities. The authors collect raw data from accelerometers and GPS using cell phones. This is pre-processed, the accelerometer data is analyzed using root mean square, threshold values for the z-axis are established, and k-means applied.

Artificial Neural Network: The structure of the neural network (NN) and the learning technique are the most important characteristics of ANN. Weighted connections between neurons are used to create a NN. An IP layer is a NN where the IP variables are entered and there are one or more secret layers among those layers. The network is termed an FF-NN when the neuron connections do not loop, which is the most basic form of ANN [84].

Deep Learning: Deep Learning (DL) refers to a set of strong reinforcement learning techniques that can handle huge amounts of unstructured data. DL methods are well suited to processing large amounts of data and computationally intensive tasks such as picture pattern identification, voice recognition, and synthesis, among others. Powerful GPUs are increasingly being utilized to conduct DL tasks as the need for CPU power grows. Deep neural networks, like ANNs, are built via DL. The term 'deep' alludes to the neural network's vast number of hidden layers. In a DL method, the number of layers is proportional to the number of computed structures. The features are automatically calculated in DL; thus, no feature computation or extraction is required before using the technique. With the progression of DL, a wide diversity of network topologies are also presented [85].

The authors train an inception NN to assist picture organization in a study on traffic accident hotspots and automated classification and detection [86]. FEDRO reveals a link between accident incidence and the actual accident location. To identify accident-prone regions, Google Maps pictures are coupled with FEDRO data to train an inception NN (TensorFlow Inception v3). The accuracy of the findings is 30%, confirming the location–accident connection theory. The inception network is a more sophisticated version of a CNN that aims to be less computationally intensive. The aim is to conduct convolution of the input data using different filters. An inception network is formed by stacking inception layers. There are numerous auxiliary classifiers in the inception network structure. [87].

Instance based: Instance-based algorithms (IBA) conduct data categorization by comparing fresh test data to training instances directly. A similarity function is used to do the comparison, and the result is given into a cataloging function [88].

Regression analysis: Regression analysis category includes methods that attempt to build mathematical models that can explain or identify the connection between two or more variables. The output of the system is the dependent. The relationships between the dependent variable and the independent variables (regressors) may be found by fitting a regression model within the system to calculate traffic congestion versus forecasts. A DT, an SVM, an MLP, and an RF algorithm were all compared to logistic regression in a study, and because the input data is time dependent, logistic regression significantly outperformed the other techniques, achieving 100% precision, 99.5% recall, and 99.9% accuracy. Furthermore, the authors claim that the LR minimal complexity allows it to be performed on low-end devices. The connection between a dependent variable and one or more independent variables is described using LR, which is a kind of regression analysis. The LR is appropriate for prediction problems in ML since it is regression analysis.

Non-Probabilistic Linear Classification: A probabilistic or non-probabilistic method may be used depending on the kind of classification issue. Non-probabilistic techniques are typically chosen when probabilities do not seem to play a role in data categorization. Other factors to consider, apart from the issue type, lead to a non-probabilistic resolution. When working with structured data, non-probabilistic categorization is helpful. Furthermore, cluster formation required in the organization is an essential factor to consider [89].

9.5.2 IoT and ML Applications in Smart Transportation

This section summarizes the six main types of smart transportation problems that are identified and how IoT technology and ML techniques are used to solve the issues. All of the studies in the six categories indicate whether or not a ML method was employed in the study. The techniques that are supported by ML are thoroughly described in the preceding section. These techniques will simply be mentioned in the appropriate category for completeness and to prevent duplication, and we will expand on the non-ML alternatives [90].

- **Route Optimization**

  Congestion is a frequent problem in cities, and it is only growing worse as the number of vehicles on the highway grows. Route optimization is a technique for recommending the most efficient route to a given location to reduce traffic congestion. Travel time and vehicle emissions are both reduced when traffic congestion is alleviated [67] Various ML methods over the IoT infrastructure have been extensively challenged in the literature to solve the route optimization issue. An MDP algorithm and V2V communication were utilized in [88, 91] for group routing to reduce traffic congestion.

- **Parking**

  Parking apps are designed to efficiently monitor parking availability, provide users with some choices, and even provide parking observation and warning methods. Many IoT devices have been utilized for this application. In addition, several studies utilize picture data to identify free parking spots in large numbers using ML algorithms.

  Amanto et al. [92] utilized CNNs to identify free parking spaces using data from smart cameras in parking lots, and also use a combination of MRF and SVM algorithms to analyze parking lot pictures. Finally, in [93], the PKLot dataset is generated, which is then evaluated using an SVM algorithm. Saarika et al. [94] suggested an intelligent parking strategy that includes an IoT-enabled parking system and a smart signpost that relays pertinent details. The parking system will utilize ultrasonic sensors to determine parking spot availability, with data collected and sent via a cloud server through a Wi-Fi module. The signpost is made from an LCD or LED panel that is powered by a Raspberry Pi and will gather and show data such as weather conditions, parking availability, distance to certain locations, and so on.

  Ultrasonic sensors at each parking space are also used to determine accessibility. The device is linked to an Arduino Uno with an ESP8266-01 Wi-Fi module and oversees transmitting data to a cloud server. The MQTT protocol is used for communication. The cloud server runs Thing Speak, an IoT platform that provides customers with a range of administration and monitoring features. Finally, end customers may download an Android app that allows them to book spaces and pay for them automatically. Gupta et al. [95] presented an end-to-end intelligent parking solution which includes a modular cloud framework, a cellular app, and a third-party monetary facility make up this system, which includes physical sensors and microcontrollers at parking spaces, as well as a modular cloud server, a smartphone application, and third-party payment service. Geomagnetic vehicle detectors analyze locations of cars at parking spaces then transmit data to the framework through a BC95-B5 NB-IoT module. There are many modules on the cloud server [96].

- **Lights**
  Smart street lights are a vital component of a smart city and are regarded as part of smart transportation services (SSL). Smart illuminations may save energy while also providing dynamic operation and management. Jia et al. [97] implemented an SSL application based on IoT infrastructure. Basic devices are all added to streetlights to make them smart. As a result, the lights can detect busy places and adjust their light level dynamically, making densely populated locations safer while also conserving energy. GPS can assist a central system in monitoring lights, their position, and condition, as well as speeding up repairs.

  The NB-IoT network is used to communicate between the SSL and the management system. The managing system is built on fog nodes, which gather information via a group of lights and periodically verify their status. Smart processes that SSLs offer may also be controlled remotely through the managing platform that has been installed. Kokilavani et al. [98] presented a comparable but simpler method of smart lighting. A Raspberry Pi is utilized as the microcontroller in this project, which connects the lamp to a light instrument, an IR LED, and an IR sensor. When the sun rises and sets, on/off signals are sent to the lamp, detected by light sensors. In addition, the lamps can detect passing vehicles or people and turn the lights on and off dynamically to save energy.

- **Accident Prevention/Detection**
  Accident prevention and detection is an intelligent transportation field and a critical movement for an effective preventive technique that may help save human lives. Road collisions may be avoided if drivers maintain a higher level of concentration during the journey. A crash prevention system may alert the driver to potentially dangerous circumstances and enable them to react quickly. By detecting accident-prone regions that have occurred in the live traffic network, accident detection may also help to decrease the number of accidents and traffic congestion. ML has proved to be very helpful in identifying road accidents or recognizing trends that could lead to future accidents and alerting drivers to prevent them.

- **Road Anomalies Detection**
  Because road conditions have an immediate effect on many elements of transportation, road anomaly discovery plays a significant role in smart transportation. A road anomaly discovery system's primary goal is to identify potholes and bumps on the street and alert drivers. Vehicle damage, traffic congestion, and road accidents may all be caused by poor road conditions. Because identifying road abnormalities is a job that lends itself to ML methods, the research covered here also takes that path. The information in this study is gathered using the k-means technique, and accelerometer data from mobile devices is analyzed. Finally, the authors compare SVM, k-NN, and RF for the job of detecting street irregularities in [99].

- **Infrastructure**
  In many ways, advances in IoT machinery have promoted current transportation, generating new ways of thinking as well as new applications for improved mobility. Changing the organization of intelligent transport systems significantly improves their competence.

The term 'social internet of vehicles' (SIoV) has been coined by combining social network concepts with the IoT for smart transportation applications. To minimize communication congestion in SIoV, Jain et al. [71] suggested a cross-layered VSNP. To speed up communication, the protocol covers the physical and network levels, and MAC. In a ring-type routing, the MAC layer separates circular time slots. WSN nodes make up the physical layer, while the network layer facilitates static access points. In simulations performed in MATLAB, the proposed protocol outperforms an existing protocol (MERLIN).

9.6 Conclusion

The chapter has illustrated that an IoT device may be reliably recognized based on network traffic parameters, improving understanding of state-of-the-art traffic management technologies, especially using data mining and ML. Ad-hoc and wireless sensor networks are also discussed. In addition, the architecture of edge computing is examined. Edge Computing minimizes network flow and hence reduces bandwidth needs in IoT because the IoT offer a plethora of new possibilities in the real world.

References

1. Jararweh, Yaser, Mahmoud Al-Ayyoub, Elhadj Benkhelifa, Mladen Vouk, and Andy Rindos. "SDIoT: a software-defined based internet of things framework." *Journal of Ambient Intelligence and Humanized Computing* 6, no. 4 (2015): 453–461.
2. Gubbi, Jayavardhana, Rajkumar Buyya, Slaven Marusic, and Marimuthu Palaniswami. "Internet of Things (IoT): A vision, architectural elements, and future directions." *Future Generation Computer Systems* 29, no. 7 (2013): 1645–1660.
3. Akyildiz, Ian F., Ahyoung Lee, Pu Wang, Min Luo, and Wu Chou. "Research challenges for traffic engineering in software-defined networks." *IEEE Network* 30, no. 3 (2016): 52–58.
4. Akyildiz, Ian F., Ahyoung Lee, Pu Wang, Min Luo, and Wu Chou. "A roadmap for traffic engineering in SDN-OpenFlow networks." *Computer Networks* 71 (2014): 1–30.
5. Liu, Jiaqiang, Yong Li, Min Chen, Wenxia Dong, and Depeng Jin. "Software-defined internet of things for smart urban sensing." *IEEE Communications Magazine* 53, no. 9 (2015): 55–63.
6. Kuang, Liwei, Laurence T. Yang, and Kai Qiu. "Tensor-based software-defined internet of things." *IEEE Wireless Communications* 23, no. 5 (2016): 84–89.
7. Erickson, David. "The beacon OpenFlow controller." In *Proceedings of the Second ACM SIGCOMM Workshop on Hot Topics in Software-Defined Networking*, pp. 13–18. 2013.
8. Chen, Yun, Weihong Chen, Yao Hu, Lianming Zhang, and Yehua Wei. "Dynamic load balancing for software-defined data center networks." In *International Conference on Collaborative Computing: Networking, Applications and Worksharing*, pp. 286–301. Springer, Cham, 2016.
9. Zhang, Lianming, Qian Deng, Yiheng Su, and Yao Hu. "A box-covering-based routing algorithm for large-scale SDNs." *IEEE Access* 5 (2017): 4048–4056.
10. Maity, Prasenjit, Sandeep Saxena, Shashank Srivastava, Kshira Sagar Sahoo, Ashok Kumar Pradhan, and Neeraj Kumar. "An effective probabilistic technique for DDoS detection in OpenFlow controller." *IEEE Systems Journal* 16 (2021): 1345–1354.
11. Rout, Suchismita, Kshira Sagar Sahoo, Sudhansu Sekhar Patra, Bibhudatta Sahoo, and Deepak Puthal. "Energy Efficiency in Software-Defined Networking: A Survey." *SN Computer Science* 2, no. 4 (2021): 1–15.

12. Singh, Simar Preet, Anand Nayyar, Rajesh Kumar, and Anju Sharma. "Fog computing: from architecture to edge computing and big data processing." *The Journal of Supercomputing* 75, no. 4 (2019): 2070–2105.

13. Kaur, Avinash, Parminder Singh, and Anand Nayyar. "Fog Computing: Building a Road to IoT with Fog Analytics." In *Fog Data Analytics for IoT Applications*, pp. 59–78. Springer, Singapore, 2020.

14. Zhong, Xiaoxun, Lianming Zhang, and Yehua Wei. "Dynamic load-balancing vertical control for a large-scale software-defined Internet of Things." *IEEE Access* 7 (2019): 140769–140780.

15. Godway, Pranav, R. Gowrishankar, Vikram SeshaSai, Venkata Sai Surya Laxman Rao Bellala, Y. Sai Kiran Kumar Reddy, Siva Sankara Sai Sanagapati, and B. V. Avinash. "Smart traffic management system based on software-defined internet of things architecture." In *2019 IEEE International Conference on Advanced Networks and Telecommunications Systems (ANTS)*, pp. 1–5. IEEE, 2019.

16. Krishnamurthi, Rajalakshmi, Anand Nayyar, and Arun Solanki. "Innovation Opportunities through the Internet of Things (IoT) for Smart Cities." In *Green and Smart Technologies for Smart Cities*, pp. 261–292. CRC Press, 2019.

17. Nayak, Rajendra Prasad, Srinivas Sethi, Sourav Kumar Bhoi, Kshira Sagar Sahoo, Nz Jhanjhi, Thamer A. Tabbakh, and Zahrah A. Almusaylim. "TBDDosa-MD: Trust-based DDoS misbehave detection approach in software-defined vehicular network (SDVN)." *CMC-Computers Materials & Continua*, 69, no. 3 (2021): 3513–3529.

18. Yan, Lu, Yan Zhang, Laurence T. Yang, and Huansheng Ning, eds. *The Internet of Things: From RFID to the Next Generation Pervasive Networked Systems*. CRC Press, 2008.

19. Palit, Achinta K. "Internet of Things (IoT) architecture—A review." In *Proceedings of International Conference on Recent Trends in Machine Learning, IoT, Smart Cities, and Applications*, pp. 67–72. Springer, Singapore, 2021.

20. Vangelista, Lorenzo, Andrea Zanella, and Michele Zorzi. "Long-Range IoT Technologies: The Dawn of LoRa™." In *Future Access Enablers of Ubiquitous and Intelligent Infrastructures*, pp. 51–58. Springer, Cham, 2015.

21. Hashem, Ibrahim Abaker Targio, Victor Chang, Nor Badrul Anuar, Kayode Adewole, Ibrar Yaqoob, Abdullah Gani, Ejaz Ahmed, and Haruna Chiroma. "The role of big data in a smart city." *International Journal of Information Management* 36, no. 5 (2016): 748–758.

22. Naik, Nitin. "Choice of effective messaging protocols for IoT systems: MQTT, CoAP, AMQP and HTTP." In *2017 IEEE international systems engineering symposium (ISSE)*, pp. 1–7. IEEE, 2017.

23. Satyanarayanan, Mahadev. "The emergence of edge computing." *Computer* 50, no. 1 (2017): 30–39.

24. Luan, Tom H., Longxiang Gao, Zhi Li, Yang Xiang, Guiyi Wei, and Limin Sun. "Fog computing: Focusing on mobile users at the edge." arXiv preprint arXiv:1502.01815 (2015).

25. Nayyar, A., Rameshwar, R. U. D. R. A., and Solanki, A. "Internet of Things (IoT) and the Digital Business Environment: A Standpoint Inclusive Cyber Space, Cyber Crimes, and Cybersecurity." In *The Evolution of Business in the Cyber Age*, vol. 10, 2020, 9780429276484-6.

26. Talari, S.; Shafie-Khah, M.; Siano, P.; Loia, V.; Tommasetti, A.; Catalão, J. A review of smart cities based on the internet of things concept. *Energies* 2017, 10, 421.

27. NIST: National Institute of Standards and Technology n.d. Available from: https://www.nist.gov/

28. Raj, P., and A. C. Raman. *The Internet of Things: Enabling Technologies, Platforms, and Use Cases*. Auerbach Publications 2017.

29. Patel, Keyur K., and Sunil M. Patel. "Internet of things-IOT: definition, characteristics, architecture, enabling technologies, application & future challenges." *International Journal of Engineering Science and Computing* 6, no. 5 (2016).

30. Mufti, T., N. Sami, and S. S. Sohail. "A review paper on Internet of Things (IoT)." *Indian Journal of Applied Research* 9, no. 8 (2019): 27.

31. Asghari, Parvaneh, Amir Masoud Rahmani, and Hamid Haj Seyyed Javadi. "Internet of Things applications: A systematic review." *Computer Networks* 148 (2019): 241–261.

32. Lakshmipriya, S. "Exploring the nuances of Internet of Things in health care assisting system." *Perspectives* 6, no. 02 (2019).

33. Atlam, Hany Fathy, Robert Walters, and Gary Wills. "Internet of things: state-of-the-art, challenges, applications, and open issues." *International Journal of Intelligent Computing Research (IJICR)* 9, no. 3 (2018): 928–938.

34. Solanki, Arun, and Anand Nayyar. "Green Internet of Things (G-IoT): ICT Technologies, Principles, Applications, Projects, and Challenges." In *Handbook of Research on Big Data and the IoT*, pp. 379–405. IGI Global, 2019.

35. Vignesh, R., and A. Samydurai. "Security on the internet of things (IoT) with challenges and countermeasures." *International Journal of Engineering Development and Research, IJEDR* 5, no. 1 (2017): 417–423.

36. Hafedh Chourabi, J. Ramon Gil-Garcia, Theresa A. Pardo, Taewoo Nam, Sehl Mellouli, Hans Jochen Scholl, Shawn Walker, and Karine Nahon, "Understanding smart cities: An integrative framework." In *45th Hawaii International Conference on System Sciences*, pp. 2289–2297, 2012.

37. Atzori, Luigi, Antonio Iera, and Giacomo Morabito. "The Internet of Things: A survey." *Computer Networks* 54 (2010): 2787–2805.

38. http://arduino.cc/en/ArduinoCertified/IntelGalileo

39. "Traffic signaling Timing Manual." Publication Number: FHWA-HOP-08-024, June 2008.

40. http://www.raspberrypi.org

41. Annual Urban Mobility Report, Texas A&M University, http://mobility.tamu.edu/ums

42. Vincent, R. A., A. I. Mitchell, and D. I. Robertson. *User Guide to TRANSYT Version 8*, JHK & Associates, 1980.

43. Cai, Chen. "Adaptive Traffic Signal Control Using Approximate Dynamic Programming." Ph.D. dissertation, University College London, 2009.

44. Kumar, Adarsh, Rajalakshmi Krishnamurthi, Anand Nayyar, Ashish Kr Luhach, Mohammad S. Khan, and Anuraj Singh. "A novel Software-Defined Drone Network (SDDN)-based collision avoidance strategies for on-road traffic monitoring and management." *Vehicular Communications* 28 (2021): 100313.

45. Ng, Kok Mun, Mamun Bin Ibne Reaz, Mohd Alauddin Mohd Ali, and Tae Guy Chang. "A brief survey on advances of control and intelligent systems methods for traffic-responsive control of urban networks." *Technical Gazette* 20, no. 3 (2013): 555–562.

46. Aldukali Salem I. Almselati, Riza Atiq O. K. Rahmat, Othman Jaafar, "An overview of urban transport in Malaysia." *Medwell Journals* 6, no. 1 (2011): 24–33.

47. World Bank. "World development indicators." 2015. [Online]. Available: http://databank.worldbank.org/data/reports.aspx?source=worlddevelopment- indicators. [Accessed 7 April 2016].

48. The Star. "The Klang Valley has finally arrived to be in a top spot in world business – Nation | The Star Online." 2013. [Online]. Available: http://www.thestar.com.my/news/nation/2013/01/02/the-Klang-valleyhas-finally-arrived-to-be-in-a-top-spot-in-world-business/. [Accessed 17 Apil 2016].

49. Pande, Sohan Kumar, Sanjaya Kumar Panda, Satyabrata Das, Mamoun Alazab, Kshira Sagar Sahoo, Ashish Kumar Luhach, and Anand Nayyar. "A smart cloud service management algorithm for vehicular clouds." *IEEE Transactions on Intelligent Transportation Systems* 22 (2020): 5329–5340.

50. Suruhanjaya Pengangkutan Awam Darat. "Klang Valley Rail Trans Map | Official Suruhanjaya Pengangkutan Awam Darat (S.P.A.D.) Website." 2016. [Online]. Available: http://www.spad.gov.my/klangvalley- rail-transit-map. [Accessed 17 April 2016].

51. Bhatia, Jitendra, Ridham Dave, Heta Bhayani, Sudeep Tanwar, and Anand Nayyar. "SDN-based real-time urban traffic analysis in VANET environment." *Computer Communications* 149 (2020): 162–175.

52. Thatcher, Linda, "Lester F. Wire Invents the Traffic Light – Photos and Stories—FamilySearch.org." 2013. [Online]. Available: https://familysearch.org/photos/stories/2128625. [Accessed 17 April 2016].

53. Göttlich, Simone, Michael Herty, and Ute Ziegler. "Modeling and optimizing traffic light settings in road networks." *Computers & Operations Research* 55 (2015): 36–51.
54. Yousef, Khalil M., Jamal N. Al-Karaki, and Ali Shatnawi. "Intelligent traffic light flow control system using wireless sensor networks." *Journal of Information Science and Engineering* 26 (2010): 753–768.
55. Collotta, Mario, and Giovanni Pau. "New solutions based on wireless networks for dynamic traffic lights management: A comparison between IEEE 802.15.4 and bluetooth." *Transport and Telecommunication* 16, no. 3 (2015): 224–236.
56. Fleck, Julia L., Christos G. Cassandras, and Yanfeng Geng. "Adaptive quasi-dynamic traffic light control." *IEEE Transactions on Control Systems Technology* 24 (2015): 1–13.
57. Xiaohong, Peng, Mo Zhi, and Liao Riyao. "Traffic signal control for urban trunk road based on wireless sensor network and intelligent algorithm." *International Journal on Smart Sensing and Intelligent Systems* 6, no. 1 (2013).
58. Pescaru, Dan, and Daniel-Ioan Curiac. "Ensemble based traffic light control for city zones using a reduced number of sensors." *Transportation Research Part C* 46 (2014): 261–273.
59. Zhang, Wei, Guo-Zhen Tan, Nan Ding, and Guang-Yuan Wang. "Traffic congestion evaluation and signal timing optimization based on wireless sensor networks: Issues, approaches and simulation." *Journal of Information Science and Engineering* 30 (2014): 1245–1260.
60. Thatsanavipas, K., N. Ponganunchoke, S. Mitatha, and C. Vongchumyen. "Wireless traffic light controller." *Procedia Engineering* 8 (2011): 190–194.
61. Ghena, Branden, William Beyer, Allen Hillaker, Jonathan Pevarnek, and J. Alex Halderman. "Green lights forever: analyzing the security of traffic infrastructure." In *8th USENIX Workshop on Offensive Technologies (WOOT 14)*, no. 8, 2014.
62. Mohammed, Mohssen, Muhammad Badruddin Khan, and Eihab Bashier Mohammed Bashier. *Machine Learning: Algorithms and Applications.* CRC Press, 2016.
63. Kubat, Miroslav. *An Introduction to Machine Learning.* Springer, Cham, 2017.
64. Wu, Qihui, Guoru Ding, Yuhua Xu, Shuo Feng, Zhiyong Du, Jinlong Wang, and Keping Long. "Cognitive internet of things: A new paradigm beyond connection." *IEEE Internet of Things Journal* 1, no. 2 (2014): 129–143.
65. Madden, Sam. "From databases to big data." *IEEE Internet Computing* 16, no. 3 (2012): 4–6.
66. Yang, Jiachen, Yurong Han, Yafang Wang, Bin Jiang, Zhihan Lv, and Houbing Song. "Optimization of real-time traffic network assignment based on IoT data using DBN and clustering model in the smart city." *Future Generation Computer Systems* 108 (2020): 976–986.
67. Al-Dweik, Arafat, Radu Muresan, Matthew Mayhew, and Mark Lieberman. "IoT-based multifunctional scalable real-time enhanced roadside unit for intelligent transportation systems." In *2017 IEEE 30th Canadian Conference on Electrical and Computer Engineering (CCECE)*, pp. 1–6. IEEE, 2017.
68. Sahoo, Sampa, Kshira Sagar Sahoo, Bibhudatta Sahoo, and Amir H. Gandomi. "An auction based edge resource allocation mechanism for IoT-enabled smart cities." *2020 IEEE Symposium Series on Computational Intelligence (SSCI).* IEEE, 2020.
69. Wu, Qi, Chingchun Huang, Shih-yu Wang, Wei-chen Chiu, and Tsuhan Chen. "Robust parking space detection considering inter-space correlation." In *2007 IEEE International Conference on Multimedia and Expo,* pp. 659–662. IEEE, 2007.
70. Araújo, Anderson, Rubem Kalebe, Gustavo Girao, Kayo Goncalves, Alberto Melo, and Bianor Neto. "IoT-based smart parking for smart cities." In *2017 IEEE First Summer School on Smart Cities (S3C),* pp. 31–36. IEEE, 2017.
71. Jain, Bindiya, Gursewak Brar, Jyoteesh Malhotra, Shalli Rani, and Syed Hassan Ahmed. "A cross-layer protocol for traffic management in Social Internet of Vehicles." *Future Generation Computer Systems* 82 (2018): 707–714.
72. Zhou, Z. H. *Ensemble Methods: Foundations and algorithms.* Chapman & Hall/CRC Machine Learning, 2012.
73. Ghosh, Arnab, Tania Chatterjee, Sunny Samanta, Jayanta Aich, and Sandip Roy. "Distracted driving: A novel approach towards accident prevention." *Journal of Advanced Computer Science & Technology* 10, no. 8 (2017): 2693–2705.

74. Lienhart, Rainer, and Jochen Maydt. "An extended set of haar-like features for rapid object detection." In *Proceedings International Conference on Image Processing*, vol. 1, pp. I-I. IEEE, 2002.

75. Sahu, Santosh Kumar, Durga Prasad Mohapatra, Jitendra Kumar Rout, Kshira Sagar Sahoo, and Ashish Kr Luhach. "An ensemble-based scalable approach for intrusion detection using big data framework." *Big Data* 9, no. 4 (2021): 303–321.

76. Friedman, Nir, Dan Geiger, and Moises Goldszmidt. "Bayesian network classifiers " *Machine Learning* 29, no. 2 (1997): 131–163.

77. Rabiner, Lawrence R. "A tutorial on hidden Markov models and selected applications in speech recognition." *Readings in Speech Recognition* (1990): 267–296.

78. Liu, Wei, Seong-Woo Kim, Katarzyna Marczuk, and Marcelo H. Ang. "Vehicle motion intention reasoning using cooperative perception on urban road." In *17th International IEEE Conference on Intelligent Transportation Systems (ITSC)*, pp. 424–430. IEEE, 2014.

79. Devi, S., and T. Neetha. Machine Learning-Based traffic Congestion Prediction in an IoT Based Smart City, *International Research Journal of Engineering and Technology*, vol.04: 3442–3445, 2017.

80. Hou, Yi, Praveen Edara, and Carlos Sun. "Traffic flow forecasting for urban work zones." *IEEE transactions on Intelligent Transportation Systems* 16, no. 4 (2014): 1761–1770.

81. Al Mamun, Mohd Abdullah, Jinat Afroj Puspo, and Amit Kumar Das. "An intelligent smart-phone-based approach using IoT for ensuring safe driving." In *2017 International Conference on Electrical Engineering and Computer Science (ICECOS)*, pp. 217–223. IEEE, 2017.

82. Ghadge, Manjusha, Dheeraj Pandey, and Dhananjay Kalbande. "Machine learning approach for predicting bumps on road." In *2015 International Conference on Applied and Theoretical Computing and Communication Technology (iCATccT)*, pp. 481–485. IEEE, 2015.

83. Sennan, Sankar, Somula Ramasubbareddy, Sathiyabhama Balasubramaniyam, Anand Nayyar, Chaker Abdelaziz Kerrache, and Muhammad Bilal. "MADCR: Mobility aware dynamic clustering-based routing protocol in the Internet of Vehicles." *China Communications* 18, no. 7 (2021): 69–85.

84. Yang, Guangyu Robert, and Xiao-Jing Wang. "Artificial neural networks for neuroscientists: a primer." *Neuron* 107, no. 6 (2020): 1048–1070.

85. LeCun, Yann, Yoshua Bengio, and Geoffrey Hinton. "Deep learning." *Nature* 521, no. 7553 (2015): 436–444.

86. Ryder, Benjamin, and Felix Wortmann. "Autonomously detecting and classifying traffic accident hotspots." In *Proceedings of the 2017 ACM International Joint Conference on Pervasive and Ubiquitous Computing and Proceedings of the 2017 ACM International Symposium on Wearable Computers*, pp. 365–370, 2017.

87. Szegedy, Christian, Wei Liu, Yangqing Jia, Pierre Sermanet, Scott Reed, Dragomir Anguelov, Dumitru Erhan, Vincent Vanhoucke, and Andrew Rabinovich. "Going deeper with convolutions." In *Proceedings of the IEEE Conference on Computer Vision and Pattern Recognition*, pp. 1–9, 2015.

88. Aha, David W., Dennis Kibler, and Marc K. Albert. "Instance-based learning algorithms." *Machine Learning* 6, no. 1 (1991): 37–66.

89. Gower, John C., and Gavin JS Ross. "Non-Probabilistic Classification." In *Advances in Data Science and Classification*, pp. 21–28. Springer, Berlin, Heidelberg, 1998.

90. Jain, Amit, and Anand Nayyar. "Machine learning and its applicability in networking." In *New Age Analytics*, pp. 57–79. Apple Academic Press, 2020.

91. Sang, Koh Song, Bo Zhou, Po Yang, and Zaili Yang. "Study of group route optimization for IoT enabled urban transportation network." In *2017 IEEE International Conference on Internet of Things (iThings) and IEEE Green Computing and Communications (GreenCom) and IEEE Cyber, Physical and Social Computing (CPSCom), and IEEE Smart Data (SmartData)*, pp. 888–893. IEEE, 2017.

92. Amato, Giuseppe, Fabio Carrara, Fabrizio Falchi, Claudio Gennaro, Carlo Meghini, and Claudio Vairo. "Deep learning for decentralized parking lot occupancy detection." *Expert Systems with Applications* 72 (2017): 327–334.

93. De Almeida, Paulo R. L., Luiz S. Oliveira, Alceu S. Britto Jr, Eunelson J. Silva Jr, and Alessandro L. Koerich. "PKLot–A robust dataset for parking lot classification." *Expert Systems with Applications* 42, no. 11 (2015): 4937–4949.
94. Saarika, P. S., K. Sandhya, and T. Sudha. "Smart transportation system using IoT." In *2017 International Conference On Smart Technologies For Smart Nation (SmartTechCon)*, pp. 1104–1107. IEEE, 2017.
95. Gupta, Aniket, Sujata Kulkarni, Vaibhavi Jathar, Ved Sharma, and Naman Jain. "Smart car parking management system using IoT." *American Journal of Science Engineering and Technology* 2, no. 4 (2017): 112–119.
96. Rizvi, Syed R., Susan Zehra, and Stephan Olariu. "Aspire: An agent-oriented smart parking recommendation system for smart cities." *IEEE Intelligent Transportation Systems Magazine* 11, no. 4 (2018): 48–61.
97. Jia, Gangyong, Guangjie Han, Aohan Li, and Jiaxin Du. "SSL: Smart street lamp based on fog computing for smarter cities." *IEEE Transactions on Industrial Informatics* 14, no. 11 (2018): 4995–5004.
98. Kokilavani, M., and A. Malathi. "Smart street lighting system using IoT." *International Journal of Advanced Research in Science and Technology* 3, no. 11 (2017): 08–11.

Index

Pages in *italics* refer to figures and **bold** refer to tables.

Printed in the United States
by Baker & Taylor Publisher Services